Micromachines

Micromachines

A New Era in Mechanical Engineering

IWAO FUJIMASA

Oxford   New York   Tokyo

OXFORD UNIVERSITY PRESS

1996

Oxford University Press, Walton Street, Oxford OX2 6DP

Oxford   New York

Athens   Auckland   Bangkok   Bombay
Calcutta   Cape Town   Dar es Salaam   Delhi
Florence   Hong Kong   Istanbul   Karachi
Kuala Lumpur   Madras   Madrid   Melbourne
Mexico City   Nairobi   Paris   Singapore
Taipei   Tokyo   Toronto

and associated companies in
Berlin   Ibadan

Oxford is a trade mark of Oxford University Press

Published in the United States
by Oxford University Press Inc., New York

The author and Oxford University Press gratefully acknowledge the financial
support provided by the Daido Life Foundation for the editing of this work.

A catalogue record for this book is available from the British Library

Library of Congress Cataloging in Publication Data

Fujimasa, Iwao, 1937–
Micromachines: a new era in mechanical engineering / Iwao Fujimasa
Includes bibliographical references

1. Mechanical engineering.   2. Nanotechnology.   I. Title.
TJ153.F86 1996
621.3–dc20   96–10918

ISBN 0 19 856513 5 (Hbk)
ISBN 0 19 856528 3 (Pbk)

Typeset by EXPO Holdings, Malaysia

Printed in Great Britain by Bookcraft Ltd,
Midsomer Norton, Avon

Preface

In the twenty-first century, biological, medical and environmental technologies will become new frontiers to maintain and improve our lives. These technologies will be supported and improved by such advanced engineering techniques as fine materials, microfabrication and new concepts of design and control methods. If we consider future medical advances, these techniques will be indispensable for developing artificial organs for life support, minimal or non-invasive measurement and therapeutic techniques, adaptable man–machine interface materials and equipment, finely detailed virtual spaces for operations and telemedicine, etc. The most important new field will be producing fine mechanisms which are guided by micromachinning technology. Micromachines are defined as machines that are constructed from micrometre-size parts and have not previously been produced.

This book summarizes 10 years of survey and research into micromachines. Today, one of main applications of micromachines is considered to be medicine. I will describe the medical applications precisely, with other industrial applications being described secondarily. Physical processes in the microscopic domain where micromachine techniques exist are thought to obey classical Newtonian dynamics. Therefore, the first stage is to try to modify the conventional millimetre fabrication techniques to microtechniques which have been developed in the fabrication of semiconductor devices and precision engineering. Many improvements have occured in microfabrication techniques during the last decade. In medicine, many tools for microscopic surgery, interventional therapy, drug delivery systems and artificial organs have already been fabricated in the micrometer domain. It is already worthwhile to use micromachine technology in these applications. I have given many possible applications in this volume. However, recently, it has become clear that the smaller machines become, the more difficult are their movements. There are many dynamic barriers which have not yet been overcome using modern techniques. Therefore, I divided the book into two parts. The first is about micromachines for traditional engineering and the second is about 'nanomachines' for future technology.

In the second part, I describe nanomachines and cells from the standpoint of biological machines. The technical approach to the subject is from large to small mechanisms (top down way), the orthodox method of mechanical engineering, and not from atoms or molecules to macro mechanisms (the bottom up way) that are more popular in molecular biology. When we observe living things, we can easily understand how biological machines such as cilia, flagella and muscles can move in spite that the size of mechanical parts are less than one micrometer. Are there any differences between microscopic and nanoscopic dynamics? In this book, you cannot find the answer to how living cells convert chemical energy to mechanical energy, but you can find some new physical phenomena that might happen in submicrometer machines. This is a new domain of mechanical engineering.

In this decade, we have begun to use many measurement equipment in the micro and nano domains with nondestructive or live mode. We can use many computer design methods. From this knowledge, we can analyse living things as machines and estimate their design principles. This is truly a big chance to find new principles to make machines which are similar to the mechanisms of living things. I hope that this book will guide a new era of mechanical engineering.

Contents

Introduction: a new challenge for mechanical engineering

1

The word 'micromachines' first appeared in my article in 1988. When I wrote it, I could not find the word micromachine specifically defined in my dictionary. Later, the National Large-scale Project of the Ministry of International Trade and Industry used it as a project name in 1989 and it is now frequently found in engineering journals.

The word is not yet strictly defined but encompasses many vague meanings. In general, it means a machine with small parts of micrometer size, but electronic and mechanical engineers may dislike using it and many expressions such as 'microelectromechanical systems', 'micromechanical systems', and 'micromachining', have appeared in journal articles—some engineers have also created words, such as micromachination and micronation. The reason why many engineers dislike the word micromachine is that they detect a popular and non-academic feel to the word. However, when we set out to create a new technological concept, we must use a simple word that includes some ambiguities and covers an interdisciplinary domain. Thus, the word has recently invaded mechanical engineering.

Why did I begin to study micromachines? Because the best example of a micro machine is life itself. Life is formed from the cell—the cell is a micromachine system in itself and is composed of microparts named organelles.

I had thought from childhood that life is a machine. Before I studied biology and medicine, I was convinced that I could understand life as a machine, but after studying biochemistry, biophysics, and molecular biology, I felt that it is a miracle and mystery that God created. Mechanics did not include an understanding of life. I approached life from the standpoint of physiological functions. I studied physiology and developed an artificial heart which looked like a mechanical pump. We could use our artificial heart for clinical trials, but it only simulates the cardiac function. We succeeded in understanding how to keep a human alive under artificial circulation, but the mechanical and structural principles of a natural heart are still unknown to us. Can we make an artificial heart with the same mechanical and structural principles as a natural heart? If we want to achieve this goal, we must understand the structure, mechanics, and energy conversion mechanism of heart muscles in terms of their function as a machine. We, as engineers, worry about difficulties which we cannot easily find answers to as there are no books written on life as a machine. We do know that the size of the basic energy conversion element is in the submicrometer domain.

Until 1990, we had no information on the mechanism. Recently, we have begun using measurement and fabrication technologies on an nanometer scale, such as scanning probe microscopes, laser confocal microscopes, laser tweezers, etc., and to connect the mechanism of life to that of micromachines.

This is a story of small machines and also a story challenging to new mechanical engineering.

2

When you start to study micromechanical systems, you might find the following three original articles cited in many references.

Gabriel *et al.* (1988) wrote about the microstructures developed by the silicon process. Electronic engineers usually called these structures microelectro-mechanical systems (MEMS or micromechatronics); mechanical engineers called them microdynamical systems or micromechanisms. The structures are integrated on a silicon wafer in batch processing methods, which are similar to the very-large-scale integration (VLSI) silicon circuits.

Drexler (1986) wrote about the molecular assembly method for human-made structures. The smallest material element that we can use as a standardized fabrication source material should be an atom. The only devices we may soon be able to fabricate with atoms might be those involving protein technology. Living organisms can produce their parts from atoms by autoassembly methods. If we want to simulate the structures of a living organism, we must develop autoassembly technology.

Feynmann (1961) indicated that it should be possible to produce new machines and a new extremely miniturized machining methodology using televised robot arms in the microscopic world.

3

MEMS: microelectromechanical systems

In early micromachine studies, much introductory work was done under contracts supported by the National Science Foundation (NSF) of the United States. The main contractors were electronic engineers who had participated in silicon planar processes. In the early stages, they developed microelec-tromechanical systems for miniaturized intelligent sensors. In 1987 and 1988, three workshops on MEMS were held at Hyannis, Massachusetts (November 12, 1987), Princeton, New Jersey (January 28 and 29, 1988), and Salt Lake City, Utah (July 27, 1987). A report of these meetings was published to generate publicity, together with a small booklet entitled *Small machines, large opportunities—a report on the emerging field of microdynamics* (Gabriel *et al.* 1988) shown in Fig. I.1.

The booklet was well titled and helped begin a new era of micromachine research and development. It is comprehensive, covering interdisciplinary technologies, and

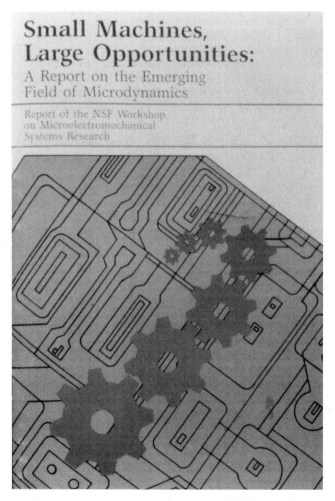

Fig. I.1 The first booklet proposing microelectromechanical systems (MEMS) published by the National Science Foundation of the USA (Gabriel *et al.* 1988). The title '*Small Machines, Large Opportunities*' was very inspiring.

explains the future possibilities of silicon micromachine technologies. Based on the report the NSF allocated 105 million dollars in 1988 and 200 million dollars in 1989 for starting MEMS research and development projects.

The report indicates that the present status of micromachine research resembles the initial stage of semiconductor research fifty years ago; the size of micromechanical parts is less than 1 mm. The applications will range from medical equipment to unmanned spacecraft and from optical communication systems to ultrahigh-density memory media.

The technique of the silicon planar process, which originates from the fabrication of semiconductor devices on a silicon wafer and was first developed at the Bell Laboratory forty years ago, has made great advances outside of micromachining technology. The photo or electron beam etching process, named lithography, is a main use of the technology, and the technique applied to fabricating a silicon wafer itself is called bulk micromachining. Depositing some layers of silicon derivatives on a silicon wafer and etching the layer by lithography is called surface micromachining. These two techniques are fundamental fabrication processes in silicon planar microfabrication technology. A first practical product was a micropressure transducer with a bending beam which was fabricated by silicon bulk micromachining. As possibilities of practical applications and fabrication techniques using silicon micromachining processes have been proved, many kinds of micromechanical systems have been proposed that could be developed using the silicon process.

Future applications of micromachines to medical, scientific, and machining instruments have already been listed in many articles, but the development of microtechnology would inevitably increase the number of new applications in the industrial field. In the future, the technologies and their products could exist in any industrial field. Moreover, fundamental research for designing and fabricating micromechanical systems required new types of interdisciplinary research work such as, microdynamical theories, material databases, designing knowledge bases, concepts on autonomic and disseminated systems, energy supply systems for microactuators and energy conversion systems, and micro–macro linkage technologies.

Gabriel *et al.* (1988) indicated the full-scale future of microelectromechanical systems.

4

Engines of creation: the coming era of nanotechnology

In a futuristic survey, we can design new conceptual micromachines using our knowledge of physics. In such a theoretical design study, which we call exploratory engineering, we can build micro- or nanomechanisms using atoms and molecules. When we can build up molecules or atoms one by one automatically, the technology will produce completely different fabrication technologies from conventional microfabrication technologies. Indeed, in protein engineering, we can actually produce molecular machines with cloned DNA coding using biosystems. K. Eric Dexler was the first explorer who tried to design an autoassembly system for a human-made machine.

The concepts of exploratory engineering have been guided by our protein engineering experience. Drexler proposed an assembler engine that would produce machine parts from atoms, molecules, and supramolecules automatically. He may be aiming to be a modern Leonardo da Vinci (1452–1519), who designed and drafted almost all today's machines in the fifteenth century. In the book *Engines of Creation*, he suggested a completely new method for producing machines using bottom-up technologies with nanosize probe systems. It resembles the methodologies of protein engineering which produce molecular machines from many kinds of amino acids using biological systems instead of mechanical probes. This shows the coming era of nanotechnologies.

Recently, in the nanoworld, the structure and dynamics of cells has gradually become clear. The size from 10 nm to 1 μm has been called the mesoscopic area and, it is in this area that almost all the mechanical principles of biological machines seem to exist. Key information to develop future new machines is hidden in the mesoscopic area.

Drexler's book indicates many futuristic technologies for machine production.

5

Plenty of room at the bottom

Richard Feynmann, Nobel Prize laureate in Physics, said at an annual meeting of American Physical Society on December 29, 1959 that there's plenty of room at the bottom. He had already described a non-biochemical approach to nanomachinary. He mentioned nanomachinary which is a method of working down, step by step, using larger machines to build smaller machines, starting from the principles of physics. When we start making machines using such methods, industrial products will have new frontiers in the micro- and nanoworlds.

Recently, we have realized the idea using the scanning tunneling microscope (STM), developed by IBM (Binnig and Rohner 1985). The instrument depends upon nanoprecise positioning technologies and it can be converted to a nanofabrication mechanism (Foster *et al.* 1988). Many material scientists and molecular electronic engineers are now engaged in atomic scale observation, tool positioning, and atom handling in ultrahigh-vacuum electron microscope chambers or molecular electron beam epitaxy instruments. The most famous product is an atom name plate; in 1990, Eiglar and Schweizer of IBM, Almaden, reported positioning single xenon atoms on a nickel crystal surface at low temperature (4 K) and used an STM to write the letters IBM using 35 xenon atoms. Thus, Feynman's forecast has already come true.

Part I
Micromachine technology

1
History of micromachines

1.1 In the USA

I shall begin by describing the short history of the micromachine project, because the concepts of the machine will then become clear.

On 9 November 1987, in Hyannis, Massachusetts, many microfabrication engineers from America, Japan, and Europe gathered together in order to discuss microstructure fabrication and its dynamics. This workshop was entitled 'Micro robots and teleoperators workshop: an investigation of micromechanical structures, actuators and sensors' and was sponsored by the IEEE Robotics and Automation Council. Many of those who pioneered micromachine technology joined the workshop—they were mainly silicon technologists. They manufactured tiny gears, turbines, electrostatic motors, and vibrators as demonstration samples. The opening address of the workshop was given by Gabriel and Trimmer of AT&T Bell Laboratories Holmdel, New Jersey. Invited talks on the subject of 'Micro structural fabrication technology' were presented with Howe of the University of California, Berkeley, in the chair. Petersen of Nova Sensor, Fremont, California, talked about the silicon micromechanics foundry; Muller of the University of California, Berkeley, reported on their microstructures, and Senturia of MIT, Cambridge, Massachusetts, discussed microrobotic devices created from silicon designing. The main techniques for such fabrication are silicon planar processes and bulk processes which have been developed in the production of integrated circuits.

Other categories of engineers were present. Some were mechanical engineers who made microrobots, such as Jacobsen of the University of Utah, Salt Lake City, Howe of the University of California, Berkeley, Fujita, of the University of Tokyo, Roppongi, Fukuda of the Science University of Tokyo, and Brooks of MIT, Cambridge, Massachusetts. These engineers all make many miniature machines without silicon technology.

Before 1987, Ehrfeld of the Kernforschungszentrum, Karlsruhe, Germany, produced many fine microsize parts using LIGA processes which were applied to their synchrotron orbital radiation equipment—he also attended the workshop. Drexler of Stanford University, Palo Alto, California, spoke about his idea of an autoassembling machine. Thus, all the stars of micromachine technology gathered for the first time. The NSF initiated two years of programmes related to micromechanical systems: the micromachine project was now beginning.

1.2 In Japan

In Japan the micromachine project started in a completely different way from the USA. In the early 1980s, Hayashi, a mechanical engineer at the Tokyo Institute of Technology (TIT) was planning to found a miniature machine society. He made many tiny mechanical actuators that were driven with electric, chemical, and pneumatic power and analysed their mechanics using scale analysis methodology. On 28 August 1988, in the University of Tokyo at Komaba, he and some university engineers and scientists held a small meeting in order to establish a micromachine research complex. The researchers' specialties were diverse, including mechanical engineering, electronic engineering, clinical medicine, robotics, material science, biosensing, and physics. They had many technological demands and dreams for micromachines. They began research on the present

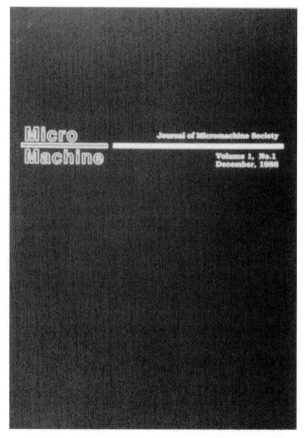

Fig. 1.1 The first issue of the journal *Micromachine* published in 1988. This might be the first appearance of the word 'micromachine' in the engineering field.

status of microfabrication technology and on its fields of application. In December 1988, the research complex (who had named themselves the Micromachine Society) held the First Symposium on Micromachines in Tokyo and the first journal devoted to micromachines, named 'Micromachine' (Fig. 1.1), was begun. Fujimasa, Nakajima, and Fujita of the University of Tokyo and Esashi of Tohoku University, Sendai, were the founders of the Micromachine Society, which aimed to make a survey of the applications of micromachines. Because the research focused on object-oriented technology, many interdisciplinary researchers and developers joined the society.

In the industrial field, since 1973 Igarashi at Toyota Central Research and Development Laboratories, Aichi, Japan, had developed micropressure sensors and accelerators combined with integrated circuits on silicon wafers. He applied these to sensing systems mounted on automobiles and to medical measurement apparatus. He joined the Micromachine Society and has promoted many conferences on micromachines (Mastsuo 1988, 1989; Harashima 1990). His research was, however, reported mainly in sensor-related societies.

In March 1989, the research complex produced a status report of micromachine technology which was supported by the Ministry of International Trade and Industry (MITI) of Japan. The report included the present technological status, fields of application, future plans, and a proposal to found a national large-scale project for micromachine research and development. Following the report, MITI founded such a large-scale national project in 1991, and the research and development of micromachines became a new engineering target in Japan. Japanese planners decided that the applications should be targeted at four categories: microelectromechanical systems, microrobots, micromachine applications in clinical medicine, and biotechnology.

1.3 In Europe and elsewhere

Practically speaking, European microfabrication technology was ahead of its time. In particular, the German government had already started a ten-year microfabrication machinery project in 1988. Heuberger and Benecke at the Fraunhofer-Institut für Mikrostrukturtechnik (IMT), Berlin, Reichel at the Forschungsschwerpunkt Technologien der Mikroperipherik, Technische Universität (TU) Berlin, and Ehrfeld at the Institut für Kernverfahrenstechnik, Kernforschungzentrum, Karlsruhe, have been key researchers into microfabrication technology. The Ministry for Research and Technology (BMFT) has funded a programme in 'microsystem technology' costing 400 million DM over the period 1990 to 1993. International Conferences on Micro-, Electro-, and Optomechanical Systems and Components were held in Berlin in 1988 (Heuberger 1989; see Fig. 1.2), in 1990, and in 1992 (Reich). Various lithographical techniques have been tried, such as wet etching (Heuberger 1989 pp. 83–124), ion etching (Pelka

Fig. 1.2 *Micromechanik* the first book about micromachines. The book discussed almost all the fundamental micromachining techniques.

and Weigman 1989), laser-assisted etching (Petzold 1989), and X-ray lithography (Huber and Betz 1989). Many micromechanical sensors have been made by Benecke (see Heuberger 1989), also microstructures (Ehrfeld *et al.* 1987; Huber and Betz 1989), optoelectronic elements (Deimel 1989), and fluidic elements. The most interesting development was the fabrication technology of X-ray lithography using synchrotron radiation. Factories such as Messerschmitt-Bölkow-Blohm (MBB) in Munich, Mercedes-Benz AG in Sindelfingen, and Siemens Nixdorf Informationssysteme AG have been built in joint ventures to apply such products to industrial equipment.

The most practical micromachine products are microelectromechanical sensors. Middelhoek at Delft University of Technology, The Netherlands, has developed microelectromechanical sensors since 1986 for radiant, mechanical, thermal, magnetic, and chemical signals using silicon technology. Micromachine research has also been announced by the University of Twente in The Netherlands and the Swiss Federal Institute of Technology, Zurich (ETH). De Rooij at the Institute de Microtechnique, Université de Neuchâtel, Switzerland has developed many microtip catheter and needle sensors, including microelectromechanical elements fabricated by silicon processes and combined electronic circuits on the sensor chip.

In Canada, the government of British Columbia made preparations for a national micromachine project at Simon Frazer University and convened an international symposium there in 1990. In Australia, the prime minister's Science and Engineering Council report of 1992 announced that a national project 'microengineering and micromachines' would start in 1993 and be joined with the Japanese national project.

Thus, micromachine research and development has taken off in the early 1990s with national and international projects organized in the main industrialized countries.

2
What are micromachines?

2.1 On the frontier of micromachines

One of the main reasons for developing micromachines is that there is plenty of industrial space where no machines exist. We don't yet have machines with conventional mechanical parts of size less than 1 mm. When we wish to develop new prototype machines with commercially available mechanical parts we go to mechanical and electronic parts shops to find the required elements. These parts conform to certain industrial standards. But the smallest parts, some types of screws and nuts, cannot be miniaturized to dimensious of less than 1 mm—and of course we cannot make a machine smaller than the size of its parts. Certainly, recent electronic machines and watches may contain many parts of submillimeter size but such parts cannot be bought 'off-the-shelf'. Many of the parts are specially made for a particular machine in a factory. Mass-production systems depend upon the replicability of the parts supplied to them, and therefore even though there are machines smaller than 1 mm at present, such machines are products of handicraft manufacturing. Even if the machining accuracy of some products can be guaranteed down to 1 μm, the processing machinery is usually huge and is used for handicraft manufacturing rather than mass-production.

At present there is no way of producing complicated three-dimensional machines from microparts; this is one more difficulty standing in the way of manufacturing micromachines. In order to build a micromachine we must develop some assembling tools, examples of such autoassembling techniques being the silicon batch process or the method used for protein engineering. Thus far the challenge of developing such micromachining techniques has not existed—there simply has been no demand for the manufacture of mechanical parts smaller than 1 mm.

If we define a micromachine as a machine system capable of assembling parts of micrometer size we find it difficult to visualize such a machine. Someone might imagine a microrobot in the bloodstream, as seen in a science fiction film; another may have an image of a colony of ants. But when we design a micromachine project with the goal of producing some industrially applicable products in the near future, the basic technology will be very limited. The final products will be gears, motors, and turbines of sizes approaching 10 μm, and the machine system will be a few millimeters in external diameter. Millimachines are thought to be industrially practicable.

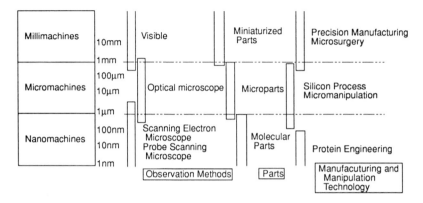

Fig. 2.1 Conventional milli-, micro-, and nano-devices.

Rather than millimachines, a developer of medical equipment usually imagines a micromachine to have nanometer size parts. A doctor must insert the micromachine system into a blood vessel or narrow cavity through a needle less than 1 mm in diameter. As the size of such a machine system would need to be less than 1 mm in external dimensions, we would need to develop the mechanical parts to less than 1 μm in size.

If a researcher intends to produce a micromachine such as a drug delivery system using a method of molecular self-assembly, their product might be of nanometer scale. Many parts inside a cell are about 1 nm in size, so when we imitate cell structure and mechanics we must start to study nanometer-scale physics.

Thus, micromachines have many factors which influence their mechanical analysis, manufacturing process, control techniques, system design, and energy source. We must build a step by step instruction manual for micromachine manufacture (Fig. 2.1).

2.2 Differences between micromachines and nanomachines

In technology, micromachines and nanomachines are classified as completely different machine categories. The differences might depend upon their production techniques and methods used to study them. In Table 2.1, fabrication methodologies are categorized into three groups: precision mechanical machining, photolithographic techniques, and atomic and molecular handling methods. The measurement instruments are also listed and these instruments are classified by the size of the objects handled.

The microtechnology necessary to support the fabrication of micromachines with movable parts, requires the development of methods to build up three-dimensional

Table 2.1 Micromachining and measuring techniques

Limit size of fabrication	Conventional machine tools	Energy beam fabrication	Etching and vapor deposition	Molecular probes	Measuring techniques
>10 μm	Conventional machine tools, vibration grinding machines	Electro-discharge milling machines	Resist film moulding, conventional dry and wet etching		Mechanical and pneumatic comparators, optical microscopes
10 μm–1 μm	Fine polishing machines, fine drilling machines,	Electro-discharge milling machines, laser assist etching	Spin coater, LPCVD, RIE, dicer, wire bonder		Fine mechanical comparators, optical and electric comparators, ultraviolet microscopes, electron spin resonance
1 μm–100 nm	Ultra fine milling and polishing machine, ultra fine forging presses, microforges		Spin coater, LPCVD, RIE, mask arrayner		Electromagnetic and electrostatic comparator, optical interferometer, phase microscopes, dark field microscopes
100 nm–10 nm	Lens polishing machines		Electron beam exposures, steppers		Laser interferometers, surface roughness meters, fluorescence microscope,
10 nm–1 nm			LIGA process, ion injection	Ion beam fabrication, molecular epitaxy	Laser confocal microscope, X-ray microanalyzer
1 nm–0.1 nm				Scanning probe microscope, protein processing system, supralattice membrane fabrication system, LB membrane fabrication system	Scanning electron microscope, scanning probe microscope (STM, AFM, etc.), electron and X-ray diffraction system, Auger, ESCA, NMR

structures in the microscopic domain. The present trend of new fabrication methods for producing micromechanical systems consists of miniaturization techniques using traditional machining and silicon process techniques. However, typical products fabricated by nanotechnology are protein molecules. Nanotechnology has inherently automatic fabrication techniques, handling molecules or atoms with nanoscale positioning in a three-dimensional field.

With regard to observation techniques, micromachines can be observed using an optical microscope and under normal atmospheric pressure, but nanomachines can only be visualized with an electron microscopic under conditions of high vacuum where no living organism can survive.

We think that animals, especially insects, are good examples of micromachines. Some micromachinists imitate mechanisms of animal locomotion for their inchworms or artificial insects. Functional biomimetics is essential knowledge for micromachine development, but the fundamental method of locomotion is on a scale of less than 1 μm. Bacteria are the smallest living organisms on the earth— no self-supporting life exists at a size of less than 1 μm. However, many movable parts and actuators exist in the submicrometer domain, i.e. cell organs or organelles. Organelles are good examples of nanomechanisms in cells or tissues, but their functional mechanisms are not yet clear. The structures of cell organelles have recently been precisely observed and their functions are gradually being revealed. The cell skeleton and motor protein are thought to be typical mechanical parts of the eukaryotic cell.

From the physical point of view, micromachine dynamics can be thought of as obeying classical Newtonian mechanics. We can estimate the dynamics using scale analysis (Trimmer 1990; Drexler 1992). But when we analyse the behaviour of mechanical parts at the nanometer level we must allow for thermal molecular motion; at least, we must use the Langevin equation for the influence of Brownian molecular motion. We cannot apply scale analysis in the nanometric domain. When we observe a biosystem as a machine and analyse its mechanics, we take into account molecular motion and the interactive forces of the molecules. When we analyse an organelle as a moving mechanical part, the mechanics should be expressed by the Langevin equation or quantum mechanics instead of bulk chemical motive forces.

2.3 Scaling laws: why micromachines have not been made

An important reason why we have not yet pursued micromachine research in the field of mechanical engineering is that the motive force is much less important in one microscopic environment than resistive forces (Fig. 2.2). Therefore, micromachines are at an inherent disadvantage in mechanical engineering. In mechanical engineering precesion of fabrication has always been pursued, but the miniaturization of parts has been relatively ignored. I have asked some famous

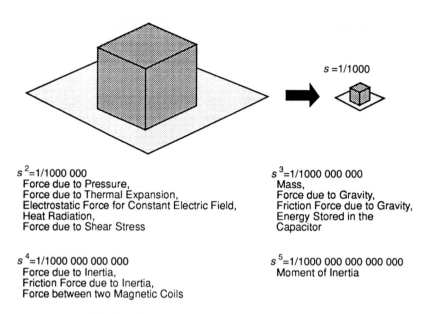

s^2=1/1000 000
 Force due to Pressure,
 Force due to Thermal Expansion,
 Electrostatic Force for Constant Electric Field,
 Heat Radiation,
 Force due to Shear Stress

s^3=1/1000 000 000
 Mass,
 Force due to Gravity,
 Friction Force due to Gravity,
 Energy Stored in the
 Capacitor

s^4=1/1000 000 000 000
 Force due to Inertia,
 Friction Force due to Inertia,
 Force between two Magnetic Coils

s^5=1/1000 000 000 000 000
 Moment of Inertia

Fig. 2.2   Geometric scaling of forces and energies.

professors of mechanical engineering why we have no incentive to make small mechanical parts, although the technology exists for making fine microparts, and the silicon process has already been applied for processing very large scale integration elements with submicrometer precision. The general opinion was that the incentive for miniaturization of *electronic* parts depends upon the improvement of electronic switching speeds. The switching speed is limited by the length of the electron transmission pathway. Thus, the size of electronic elements has inevitably decreased and as a result electronic parts have achieved micrometer size. Such an incentive does not exist for machine miniaturization.

As there has been no incentive to develop microsize parts, precise dynamic analysis of microstructures has not been attempted. From micrometer to nanometer, we have no basic design principles that are supported by quantum mechanics and/or classical mechanics. The material characteristics and mechanics of micrometer or submicrometer structures might be theoretically known in physics and chemistry, but practically there is almost no supporting knowledge to help us design micromechanical systems. Some mechanical engineers have analysed the movement and the dynamics of animals, and made some biomachines by simulation methods, a process which has been called biomimetics. But there have been no systematic or mechanical analyses concerning a cellular system composed of many micromechanical parts (Hirose 1987).

A practical bottle-neck hindering development of a microactuator is the fact that the resistive forces, such as viscous drag and surface tension, exceed the inertial

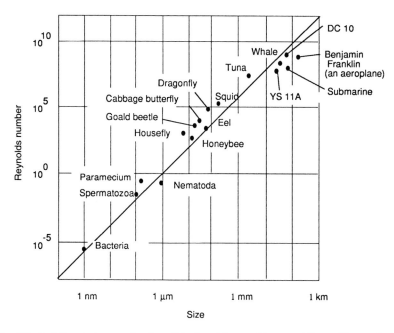

Fig. 2.3 Relation between size (length) of mobile machines and Reynolds numbers of their fluid environments (adopted from Hayashi (1988)). The Reynolds number becomes around 1 when mobile machines are around 1 mm. The Reynolds number expresses the ratio between inertial force and viscous resistance. Therefore, if the machine is less than 1 mm viscous resistance dominates the dynamics. (YS 11A is a twin-engined plane.)

and electromagnetic forces which are the typical driving forces of small motors. The disadvantage originates in the fact that the inertial forces scale as the fifth power (mass × length^2) and the viscous drag and surface tension forces scale as the second power (area). When the length of a machine decreases from 1 mm to 1 μm, the area decreases by a millionth and the volume decreases by a billionth. The resistive forces, proportional to the area, increase one thousand times more than the forces proportional to the volume. Therefore, in order to develop a microactuator we should select driving forces which scale advantageously into the micro domain. Electrostatic forces, hydraulics, and pneumatics usually scale as the dimension to the second power. I discuss the precise analysis in another chapter, but scale analysis sets strong limits on the power efficiency in the microscopic domain (Fig. 2.3).

Biologists have observed movement of cells and organelles for a long time, but they have never doubted why organelles and cells are able to move. Biomechanical systems move in water, and when we analyse biomechanics using classical Newtonian mechanics, viscous resistance dominates the total movement. As the Reynolds number of cells and organelles, which is the ratio of inertial

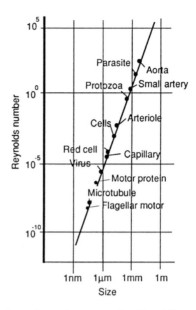

Fig. 2.4 The hydrodynamic environment of self-mobile cells in vessels. Because the size of cells and the inner diameters of vessels are less than 1 mm, the Reynolds numbers do not exceed 1. Viscosity is the most dominant factor of power loss in microvessels. However, the self-mobile cells move easily in the domain; they might use the viscous drag force itself. Actuators of cell organelles such as motor proteins and flagellar motors follow completely different dynamics to those of micrometer-sized self-mobile cells. The line in the figure extends from that in Fig. 2.3.

forces to viscous resistance, becomes 1 at 1 mm, viscous resistance is very dominant in biological machines (Fig. 2.4). In mechanical engineering terms a biological machine should not work—but it moves well. We estimate that some other mechanical principles must govern motion in living organisms. We must research the principle not only from the molecular behaviour of proteins but also the mechanics of supramolecule, which are built into a structure using many proteins and lipids.

Good examples of precise micromachines, therefore, are cells, tissues, and organs. We should survey the mechanism and functions of cells with a view to designing our own micromechanical structures. A major target of recent biotechnology has been to understand the functioning of cells in terms of molecular interaction. Instead of being a solid molecular reaction survey, micromachine research is based upon discovering the rules of design and fabrication by which the cell was constructed from protein molecules. Microbiotechnology and macro-precise engineering studies are thus complementary.

In order to design molecular machines using our knowledge of cell anatomy, we must develop micro or nano-sensing systems as there are no suitable conventional methods of dynamic physical measurement on this scale. To measure the power of a microsystem a possible method might be stress–strain relation analysis using a purpose-made microbeam, the Young's modulus of which is already known; however, we must develop the measurement system first.

In order to develop micromachines, materials research, dynamics of microsystems, and other interdisciplinary research will be required. Micromachine research will bring together scientists and engineers from many fields and will create a new interdisciplinary technology.

2.4 Dynamics of micromachines

2.4.1 Theoretical analysis

In mechanics textbooks we classify actuators into force types, that produce a certain force, and work types, that generate movement (Hayashi 1985). Usually, we combine the two types of actuators and obtain the force and displacement required for a particular application.

Force actuator

The control system of a typical force-producing actuator is shown in Fig 2.5A. When mass of a motor rotator (m_M) and a load (m_L) combine with a gear system which has the optimal gear ratio $R = \sqrt{m_L/m_M}$ and match with adequate impedance, and the system is controlled ideally as shown in the case of (a-1), we can operate the system rigid body until the limit frequency of

$$\omega < (1/3)\sqrt{k/(m_M//m_L)}.$$

If we wish to use the system beyond the frequency limit, we should put out the load from the feedback loop and the motor rotator position feeds directly back to the system. Thus, we can use the system up to the limit of the resonance frequency. In the system, force (F), amplitude (a), frequency (ω) and acceleration (α) are written in the following formulae:

$$F = 2a\omega\sqrt{m_M m_L}$$
$$\omega < \sqrt{k/m_L}$$
$$\alpha = (F/\sqrt{m_M})\,(1/2\sqrt{m_L}).$$

Displacement actuator

A piezoelectric actuator is an example of a displacement actuator. If we wish to use such an actuator with a larger displacement, we can add a displacement-enlarging mechanism such as levers or harmonic motors. When a stable displacement function is obtained we can then add some feedback control mechanisms. We can calculate the operating regions of the actuators in output

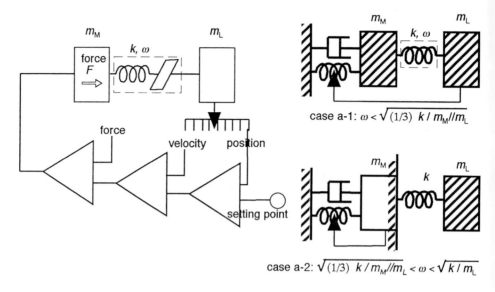

case a-1: $\omega < \sqrt{(1/3)\ k\,/\,m_M /\!/ m_L}$

case a-2: $\sqrt{(1/3)\ k\,/\,m_M /\!/ m_L} < \omega < \sqrt{k\,/\,m_L}$

Block diagram of force actuator

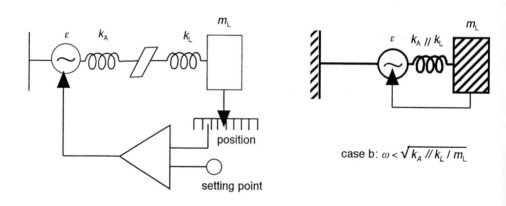

case b: $\omega < \sqrt{k_A /\!/ k_L\,/\,m_L}$

Block diagram of displacement actuator

Fig. 2.5A Characteristic analysis of force actuator (a) and displacement actuator (b), where m_M = equivalent mass of a motor rotator; m_L = mass of a load; $m_L /\!/ m_M$ = optimal gear ratio; k = coupling constant between a load and an actuator; $k_A /\!/ k_L$ = optimal coefficient of coupling between a load and an actuator; ω = vibration frequency; a = amplitude of vibration; F = force of a force actuator; ε = displacement of an actuator.

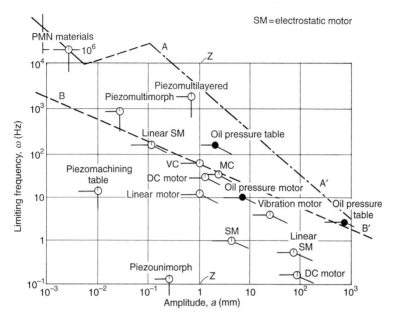

Fig. 2.5B Output amplitude (*a*) and limiting frequency (*ω*) of commercially available actuators (Hayashi 1993). The figure illustrates the following points: B–B′: electromagnetic actuators exist below the line, and some fluid actuators are used beyond the line; A–A′ solid state actuators are used with working distance less than 1 mm (left zone of line Z–Z′) and cannot exceed the line B–B′ over an amplitude of 1 mm even if we apply some enlargement mechanisms; friction motors show a good result.

amplitude (*a*) and limiting frequency (*ω*), which are calculated from

$$a = Rl\varepsilon$$
$$\omega < \sqrt{(k_A // k_L) / m_L} = \sqrt{ES / Rlm_L}$$

where R = enlargement rate, l = length of actuator, ε = displacement of unit length of actuator, $k_A // k_L$ = optimal coefficient of coupling between load and actuator, m_L = mass of an actuator, E = Young's modulus of actuator, and S = cross section area of actuator.

When we plot the amplitude against limiting frequency for many commercially available actuators we obtain the results shown in Fig. 2.5B.

2.4.2 Scaling effects on actuators

Every kind of actuator has its own force generation principles. When we make a microactuator we must analyse the relationship between its generating force and resistance. Scaling analysis is the most popular technique for assessing the effects of micronization. For example, many fundamental physical quantities have specific scaling exponents (Drexler 1992).

In a mechanical system, when the length L is reduced to 1/10 its original value, the displacement and deformation, torque, and stiffness become $L^1/10$; area, force, and strength become $L^2/100$; and mass and volume become $L^3/1000$. As acceleration is obtained from force/mass, the scale exponent becomes L^{-1} and that of vibration frequency ($\sqrt{\text{stiffness/mass}}$) is also L^{-1}. We usually think that time is not related to the scaling exponent (L^0) but when we introduce time as proportional to 1/frequency, time has the scaling exponent L^1. As speed is obtained from acceleration × time, it scales as L^0 and power, which is proportional to force × speed, scales as L^2. Mechanical power density, proportional to power/volume, scales as L^{-1} and viscous stress at constant speed is proportional to shear stress and it is calculated from speed/thickness. Thus, the viscous stress and shear stress scale as L^{-1}. One of the most important resistive forces is frictional power, which is proportional to frictional force (L^2) × speed (L^0). Therefore, frictional force and frictional power have scaling exponents L^2. As wear life is thought to be proportional to the thickness of materials/erosion rate, the scaling exponent becomes L^1.

In classical steady-state electrostatic systems the fundamental scaling exponent is electrostatic field (L^0). Voltage is obtained from electrostatic field × length, therefore voltage has the scaling exponent L^2. Electrostatic force is proportional to area (electrostatic field)^2; (L^2); resistance is proportional to length/area (L^{-1}); current is obtained from voltage/resistance (L^2). Following such reasoning we can calculate the scaling exponents of electromagnetic systems: resistance scales as L^{-1}; the capacitive time constant scales as L^0; voltage, capacitance, inductance, magnetic field, electrical oscillation frequency, and oscillator Q; scale as L^1 electrostatic force, current, and inductive time constant scale as L^2; electrostatic energy scales as L^3 magnetic force scales as L^4; magnetic force is (L^5); motor resistive power density scales as L^{-2}; and motor current density scales as L^{-1}.

In classical thermal systems, the basic assumptions about scaling exponents are based on considerations of heat capacity which is proportional to volume (L^3). Thermal conductance is derived from area/length, therefore it scales as L^1, and the thermal time constant is derived from heat capacity/thermal conductance, which means it scales as L^2; temperature change is proportional to frictional power/ thermal conductance (L^1).

A summary of scaling laws is given in Table 2.2.

2.4.3 Viscosity

We estimate that many applications of micromachines will be in air or liquid as mobile machines or microrobots. The Reynolds number of a machine of size 1 mm^3 is roughly estimated at 100. When we analyse microrobot dynamics with such a low Reynolds number, we make a geometrically similar macromodel, set it

Table 2.2 Summary of scaling laws (adapted from Drexler (1992))

Physical quantity	Scaling magnitude (L^n)	Guiding rule	Typical exponent n of a micromachine
Area (S)	2	$S \propto L^2$	10^{-12} m^2
Force (f)	2	$f \propto S$	10^{-2} N m^{-2}
Total strength (T)	2	$T \propto f$	
Stretching stiffness (St)	1	$St \propto S/L$	10^6 N m^{-1}
Share stiffness (Sh)	1	$Sh \propto St \propto L$	
Bending stiffness (Bst)	1	$Bst \propto radius^4/L^3$	
Deformation (Df)	1	$Df \propto f/St$	10^{-8} m
Mass (m)	3	$m \propto Q(volume)$	10^{-15} kg
Acceleration (a)	-1	$a \propto f/m$	10^{-15} m s^{-2}
Frequency (w)	-1	$w \propto$ acoustic speed/L	10^{13} rad s^{-1}
	-1	$w \propto \sqrt{St/m}$	
	-1	$w \propto v/L$	
Time (t)	1	$t \propto 1/w$	
Speed (v)	0	$v \propto a \times t$	10^3 m s^{-1}
Power (e)	2	$e \propto f \times v$	10^{-8} W
Power density (ed)	-1	$ed \propto e/Q$	10^{19} W m^{-3}
Share rate (t)	-1	$t \propto v/L(thickness)$	
Viscous stress (V-st)	-1	$V\text{-}st \propto t$	10^6 N m^{-2}
Frictional force (fr)	2	$fr \propto f$	
Frictional power (fr-e)	2	$fr\text{-}e \propto f \times v$	
Wear life (t-we)	1	$t\text{-}we \propto L/erosion\ rate$	
Thermal speed (t-sp)	-1.5	$t\text{-}sp \propto \sqrt{thermal\ energy/m}$	100 m s^{-1}
Electrostatic field (E)	0		
Voltage (V)	1	$V \propto E \times L$	1 V
Electrostatic force (E-f)	2	$E\text{-}f \propto S \times E^2$	10^{-12} N
Resistance (R)	-1	$R \propto L/S$	10 W
Ohmic current (A-o)	2	$A\text{-}o \propto V/R$	10^{-8} A
Field emission current (A-f)	2	$A\text{-}f \propto S$	
Electrostatic energy (E-e)	3	$E\text{-}e \propto Q \times E^2$	10^{-21} J
Capacitance (C)	1	$C \propto E\text{-}e/V^3$	10^{-20} F
Electrostatic power (E-p)	2	$E\text{-}p \propto E\text{-}f \times v$	
Electrostatic power density (E-pd)	-1	$E\text{-}pd \propto E\text{-}p/Q$	
Resistive power density (R-pd)	0	$R\text{-}pd \propto (current\ density)^2$	
Motor current density (M-d)	-1	$M\text{-}d \propto E \times w \propto (charge/S) \times w$	
Motor resistive power density (M-rd)	-2	$M\text{-}rd \propto (M\text{-}d)^2$	
Magnetic field (M)	1	$M \propto A/L(distance)$	10^{-6} T
Magnetic force (M-f)	4	$M\text{-}f \propto S \times M^2$	10^{-23} N
Magnetic energy (M-e)	5	$M\text{-}e \propto Q \times M^2$	
Inductance (I)	1	$I \propto M\text{-}e/A^2$	10^{-15} h
Inductive time constant (tc-I)	2	$tc\text{-}I \propto I/R$	$< 10^{-16}$ s

Table 2.2   (cont'd).

Physical quantity	Scaling magnitude (L^n)	Guiding rule	Typical exponent n of a micromachine
Capacitive time constant (tc-C)	0	tc-C $\propto$ R $\times$ C	
Oscillation frequency (w)	-1	w $\propto \sqrt{1/l} \times$ C	$> 10^{18}$ rad s^{-1}
	-1	w $\propto$ wave speed/L	
Quality of oscillator (Q-w)	1	Q-w $\propto$ w $\times$ I/R	
Heat capacity (Cq)	3	Ch $\propto$ Q	10^{-21} J K^{-1}
Thermal conductance (Th-cd)	1	Th-cd $\propto$ S/L	10^{-8} W K^{-1}
Thermal time constant (Th-t)	2	Th-t $\propto$ Cq/Th-cd	10^{-13} s
Temperature elevation (Th-a)	1	Th-a $\propto$ fr-e/Th-cd	

in a flow with the same Reynolds number and estimate the dynamics. But this assumption is only realistic when the system is in a stable condition (laminar flow). Usually a microrobot moves in an unstable condition (turbulent flow) and our assumption would not hold. Researchers in fluid dynamics have generally been interested in turbulent conditions with high Reynolds numbers and have not analysed precisely flows with low Reynolds numbers. A few biophysicists have analysed flow in a capillary or a small vessel. The methodologies for analysing such flow are as follows:

(1) using and calculating numerically the Navier–Stokes equation;
(2) creating a test model and making measurements on it.

A microsensor used to measure mechanical vibration is a good object for such an analysis. As a mechanism becomes smaller its resonant frequency increases. Using this principle, pressure, vibration, and chemical sensors have been developed. But the amplitude of the oscillation decreases according to the decreasing size of the mechanism because in the micro domain viscous force become more dominant than frictional force and other resistive forces. A good example of the effect of viscous force has been reported on the vibration Q analysis of a comb vibration actuator driven in air and in a vacuum. Vibration Q is a good index by which to evaluate vibration. The Q of a comb-shaped microelectrostatic oscillator vibrated in air reduces to 1/1300 of that in a vacuum (1 Pa) (Yoshimi 1992). The result indicates the strong influence of the viscosity of air on movement of a microdevice.

Analysing the movement of two plates at a constant speed (U), the space between the plates being filled with a viscous fluid, we usually apply Couette flow analysis. In Couette flow, the flow velocity (u) between the two plates is written as

$$u_y = U_y/h$$

where y is distance and h distance between the two plates. But the Q calculated from the Couette flow formula shows very high values compared with the experimental data for the microcomb actuator. The reason for the low Q is the viscous forces.

When the length of a micromachine becomes less than a few micrometers, the Reynolds number becomes nearly 1. In such a flow, the inertia effect of the Navier–Stokes equation becomes very small or negligible. We call this type of flow Navier–Stokes flow, and we can calculate more precisely the vibration Q. The Navier–Stokes formula is written as follows

$$\rho \left(D\boldsymbol{u}/Dt \right) = \rho \boldsymbol{K} - \operatorname{grad} p + \mu \Delta \boldsymbol{u}$$

where ρ is the density of the fluid, $\boldsymbol{u}$ the flow velocity, $\boldsymbol{K}$ the external forces, p the pressure and μ the coefficient of viscosity.

Lift and drag

Insects are good examples of microrobots. We can classify insect movement into three types: walking, leaping, and flying. Leaping and flying are dominant for insects less than 1 mm but flying requires the smallest energy consumption. A microrobot which simulated insect movement in fluid flow would experience lift and drag forces. Large birds and aeroplanes have large wings and the lift forces are larger than the drag forces—they fly using lift forces. But in small insects like bees and mosquitoes, drag force become large and the lift force becomes relatively small—they fly using drag forces. Therefore, a microrobot cannot glide in an air flow, and if we want to make it fly, the drag force produced by down beating should exceed that produced by up beating, as in bees or mosquitoes.

The lift force (L) and the drag force (D) are given by the following two formulae:

$$L = C_{\mathrm{L}} \frac{1}{2} \rho A u^2,$$

$$D = C_{\mathrm{D}} \frac{1}{2} \rho A u^2.$$

For the dynamic analysis experiment, a micro-aeroplane of length 1.56 mm and mass 10.8 mg was fabricated using a silicon wafer and was fixed on to a thin glass needle; the lift and drag forces were measured under parabolic laminar airflow. As the measured ratio of lift/drag becomes 0.4 (ordinary aeroplanes have a ratio of 10–100) the angle of attack falls to less than 45°. Therefore it is not possible to make a glides with a length of less than 1 mm (Kubo *et al.* 1993).

2.5 Tribology of micromachines

Surface reactions are enhanced in micromachines. As tribological properties are related to the surface and movement, it is essential to analyse and design the surface of micromachines with such properties in mind. 'Microtribology' is the interfacial reaction of surfaces on the molecular and atomic scale. From Newton's second law of motion, the inertia decreases proportionately to the volume but the atomic force and the frictional force decrease proportionately to the area. The smaller the scale of a machine becomes, the larger the relative frictional force does. When we design a control system for a micromachine we must consider the frictional force and some other nonlinear physical parameters related to surface phenomena.

In the development of micromechanical parts, many tribological effects have been observed. Early on it was discovered that the axis of a silicon micromachined turbine was worn away in a short time at 10^6 rpm rotation (Gabriel et al. 1990). The rotating speed of a micromotor was affected by friction and limited to 1/1000 of the designed value (Tai and Muller 1990). When the control system of a hill-climbing microrobot was designed to control the direction of movement, the coefficient of friction μ had to be kept constant for keeping straight forward and be changed according to the direction of motion (Hayashi 1990).

Adsorption of water on the surface of a micromachine changes the characteristics of motion markedly. This is due to the surface tension of water and the observed increased in frictional force between a magnetic disk and its head at the beginning of rotation, which is called 'sticktion'.

Many ideas have been proposed to overcome such effects in micromachines. Non-contact bearings or supporting axes were designed applying van der Waals atomic force suspension bearings (Drexler 1987) or supermagnetic floating axes and linear motors (Kim et al. 1989). Lubrication and suspension of motor proteins and cytoskeletons in a cell have been surveyed, and some proposals reported (Fujimasa 1992).

2.5.1 Solid state tribology

In frictional terms microtribology is completely different from macrotribology. In microtribology, molecular or atomic size abrasion must be considered. In micromachines the use of non-wearing contacts is thought to be the most practical approach (Table 2.3). The most important tool of analysis in the study of such contacts is the scanning probe microscope (SPM). The friction force microscope (FFM) is a variant of the atomic force microscope (AFM). The diamond tip of an AFM is in contact with and scanned on the surface with a micropressure load and the distribution of frictional force or shape of a scratched scar is measured. A silicon surface injected with nitrogen ions shows a smaller coefficient of friction than an ordinary silicon wafer; the higher acceleration voltage for ion injection causes the lower coefficient value. Frictional wear testing using an FFM shows good antifrictional characteristics in diamond fluoride, ion-injected diamond, and cubic boron nitride films (Miyake and Kaneko 1992).

Table 2.3 Microtribology for friction and wear (adapted from Miyake and Kaneko (1992)

		Microtribology	Conventional tribology
Theory and mechanism	Friction	Interaction of atoms and electrons molecular dynamics	Roughness, adhesion, plastic deformation
	Wear	Molecular dynamics (movement and falling out of atoms)	Abrasive wear, adhesive wear, wear due to plastic deformation, finite element methods
Evaluation methods	Characteristics of surface	Scanning tunneling microscope, physical analysis	Touch needle surface tester, pin on disk
	Frictional and adhesive power	Atomic force microscope, frictional force microscope	
	Wear Disk testing instruments	Microwear testing	
Anti-wear materials		Molecular theoretical zero wear: (1) small surface energy (2) high tension materials	Zero wear: (1) materials with anti-wear cracks (2) materials of less cell structures (3) small plastic flow formation (4) small adhesiveness
Applications		(1) lubricant, protection membrane (2) dust less, low cracking, surface process	(1) conventional machines (2) conventional parts

In order to obtain better lubricants in the micro domain, SPMs have been used to look for materials with a low surface energy. As fluoridation decreases the surface energy, HOPG, one of the good solid state graphite lubricants, was fluoridated by plasma processing and its surface observed by an SPM. The surface structure obtained using an FFM shows specific features indicating a low-frictional area (Bhushan *et al.* 1995).

The hardness of materials also affects their frictional wear. As C_3N_4 is thought to be theoretically harder than diamond, friction scars on nitrogen-including carbon film fabricated by an ion-plating method in nitrogen plasma were compared with silicon film and ion-plated carbon film after scratching by the diamond tip of an AFM probe. The depth of the scratched scars was measured to be 400 nm on silicon film, 50 nm on ion-plated carbon film, and 0.5 nm on nitrogen-included carbon film, after scratching three times with a pressure of

Table 2.4   Examples of zero wear in microtribology (adapted from Miyake 1992)

Characteristics of zero ware materials	Examples
No share in materials which compose friction interaction	Low surface energy
No molecular movement at frictional surface	C–F compounds
The interfacial system where always share happens	Low polilization and tight bonding
The material system defects caused by friction are negligible	Strong materials Si–C ; diamond
Tight connection to base plate	Composit fixed board Si–C/Si

70 μN. A similar methodology with various probe tip and surface materials combinations has been used to discover other antifrictional surface materials (Miyamoto *et al.* 1993). Some wear-free surface materials have been proposed and micromechanical parts fabricated using the researched materials (Table 2.4).

2.5.2 Tribology of fluid lubricants

Some micromachines have fluid lubrication systems. The micromechanical characteristics of fluid lubricants are different from those in a macro system. In microdevices the lubricant film is usually thin, sometimes reaching tri-, bi-, or monomolecular layers. As the film becomes thinner, the structure of a high-polymer lubricant influences the coefficient of friction. Monolayers of ball-shaped high-polymer lubricants, such as cyclohexane and octamethylcyclotetracyloxane, show higher coefficients of friction than bi- or trilayered lubricants, and show a very high practical viscosity. But chain-shaped high-polymer lubricants, especially those with side chains, such as polydimethylcyloxane or polybutagene, do not change coefficients of friction with the thickness of the lubricant layers (Israelachvili *et al.* 1990). The reason is that a ball-type monolayer lubricant acts as a solid state surface, which is usually called stick–slip, but a chain-type acts as an oil lubricant. This results show us that high polymers with side chains are better than ball-shaped polymers for microdevice lubrication.

Liquid crystals are liquid but have anisotropic physical characteristics. They are also used as lubricants with variable viscosity for microdevices, because their viscosity is influenced by the external electric field (Biresaw 1990). The viscosity of a liquid crystal in a field of 200–1000 V mm^{-1} increases several times more than in the absence of an electric field. Liquid crystal, can be used in a microelectric clutch of a microdevice, because the shear speed is usually very low.

3
Fabrication and manufacture

3.1 The silicon process

Micromachine research began as a spin-off technique of the silicon process. The silicon process is a practical technique for fabrication of many microparts that have a three-dimensional structure (Fig. 3.1). The process requires complicated manufacturing procedures in a clean environment but can produce many three-dimensional parts on a silicon wafer. The fabrication of large-scale integrated circuits is a good example of the process. Recently, we have been able to draw lines 0.5 μm wide on a silicon wafer using an electron beam lithographic method. The technology is far more precise than conventional machining.

Microsensors produced using the silicon process have been developed since 1980. Circuits of preamplifiers and logic elements have been integrated with transducers and used to make an intelligent chip element, called an intelligent sensor or a smart sensor. Many intelligent sensors which detect pressure, gas flow, chemical and gas concentration, etc., have been reported since the early of 1980s (Muller *et al.* 1991). Microsensors and microactuators with movable parts have been developed since 1985 and have become the first applications of micromechanical parts in the industrial field. Accelerometers and micropressure

Fig. 3.1 A crown placed on an ant's head. The crown was fabricated by a huge and highly precise conventional computer-controlled milling machine. We can produce such microproducts on an individual basis but the production cost is high and the cost/benefit ratio very low (courtesy Nippon Denso).

sensors applied for controlling vehicles are typical examples of early micromachine products. These parts have been developed mainly by the silicon process.

The size of micromechanical devices produced today ranges from 10 to 100 μm and single-crystal silicon is being employed in micromechanics because of its excellent mechanical properties and its well-established process technologies. Process techniques are classified into two categories: bulk micromachining and surface micromachining. The bulk method includes sculpting the silicon substrate using chemical etchants. The surface method involves etching several layers of thin films, such as SiO_2, Si_3N_4, etc., that have been deposited upon the substrate.

3.1.1 The bulk process

The fundamental principles of the silicon process depend on the characteristics of the silicon crystal. Since the 1950s, chemical etching of silicon has been developed using isotropic etchants such as HF, HNO_3, and CH_3COOH (Robbins and Schwartz 1958). In 1967, anisotropic etching was reported (Waggener et al. 1967) and KOH or other alkaline solutions have been used to etch different planes of silicon crystals at very different rates (Bean 1978). In particular, the (111) crystal facet of a silicon substrate can be cleanly sculpted by etching (Bassous 1978). When the surface of the silicon wafer with a specific crystal orientation is dipped in or exposed to such etchants, cavities, and grooves with precisely angled walls can be created in combination with unetched crystallographic planes or with the planes protected by several masking materials. The area of the wafer surface that is not to be etched is masked with conventional thin-film materials, such as SiO_2 or Si_3N_4 (Petersen 1982) (Fig. 3.2).

While electrochemical etching of silicon has been developed for a number of years (Turner 1958), practical applications of the technique have only been gradually realized (Branebjerg et al. 1991) If we want to design the process to stop the etch in a particular layer, the silicon layer should be doped heavily with boron or should receive an application of a passivating voltage to one side of a p–n junction. The preciseness of the microfabricated structure is determined by the crystal structure of the silicon wafer, by the etch-stop layer thickness, or by the lithographic masking pattern (Fig. 3.3).

3.1.2 The planar process

The most widely used technique for producing a moving mechanism in silicon micromechanics is surface micromachining. Sacrificial layer technology is used to create a completely batch-processed micromechanical structure. A sacrificial layer is a thin film, composed of SiO_2 or epitaxial Si, deposited in the surface micromachining process and etched away later to reveal a microstructure (Fig. 3.4). The sacrificial and structural layer films are grown on silicon wafers by

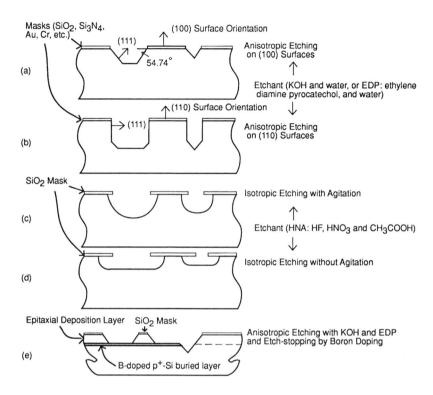

Fig. 3.2   A summary of wet chemical etching (adapted and modified from Terry *et al.* (1979) and Runyan *et al.* (1967)). In wet chemical etching KOH is frequently used for anisotropic etching. In KOH solution, the (111) surface direction of silicon crystal is etched at a very low speed compared with the (110) direction Therefore, a silicon wafer directed (100) surface can be etched by KOH etchant with 54.74° slopes (a) and a wafer with (110) surface direction can be etched around vertical slopes (b). HNA (HF+HNO$_3$+CH$_3$COOH) is commonly used as an isotropic etchant. As the etching speed does not depend upon the crystal axis, the etched pattern follows the isotropic direction (c,d). However, etching characteristics depend on the silicon dopant concentration (e), the mix ratios of the three etch components, and even the degree of etchant agitation (c). A deeply doped silicon layer can be used as an etch-stop layer.

low-pressure chemical vapour deposition (LPCVD). Such a concept was first applied with metal films in the 1960s at the Westinghouse Research Laboratories, Pittsburgh, Pennsylvania (Guldberg *et al.* 1975). The most famous and earliest application of this process to micromechanical elements is that of beams, bearings, and linkages using polycrystalline silicon with SiO$_2$ as the sacrificial material, developed in the 1980s at the University of California at Berkeley (Fan *et al.* 1988), and the Toyota Central Research and Development Laboratories, Inc., Aichi, Japan (Sugiyama *et al.* 1986) (Figs 3.5 and 3.6).

3.2 The LIGA process

Using conventional silicon processes, we cannot build the surface deposition up to a thickness of more than a few layers. Therefore, the thickness of a deposition layer cannot reach more than 10 μm and the micromechanical devices are only two-dimensional in structure. In Germany, the LIGA process has extended the range of batch-fabricated structures to truly three-dimensional structures.

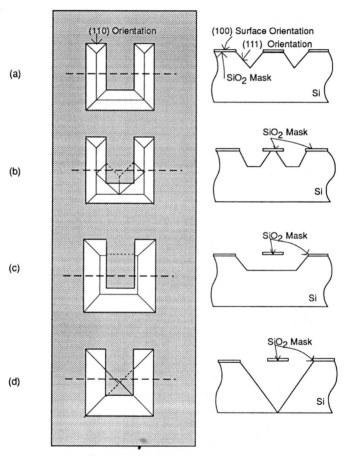

Fig. 3.3 A schematic diagram of typical pyramidal pit, bounded by the (111) planes etched into (100) silicon with anisotropic etch through a complicated silicon oxide mask. (a) Type of pit which is expected from an anisotropic etch with a slow convex undercut rate. (b) The same mask pattern can result in a substantial degree of undercutting using an etchant with a fast convex undercut rate such as EDP (ethylene diamine + pyrocatecol + water). (c) Further etching of (b) produces a cantilever beam suspended over the pit; a typical piramidal pit is formed after a sufficiently long etching time. (Adapted and modified from Petersen (1982).)

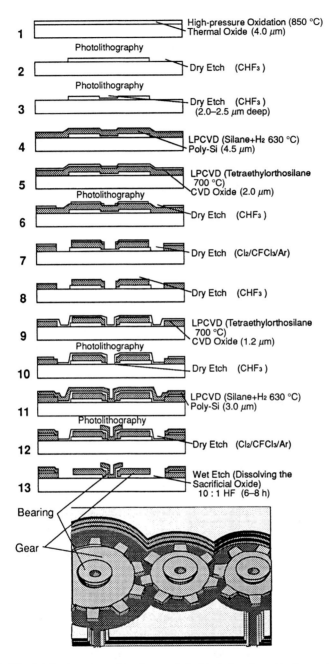

Fig. 3.4 The fabrication procedure for an air turbine, developed in the early stages of micromachine research. However, the micromachining process included almost all the modern silicon process. (This illustration of the three sets of gears is adapted from a photograph taken by Meheregany *et al.* (1988).)

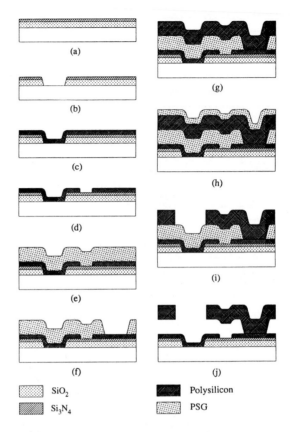

SiO₂   Polysilicon

Si₃N₄   PSG

Fig. 3.5 In order to illustrate the standard silicon planar process, we extract a fabrication protocol from Tang *et al.* (1989) who belong to Howe's group at the University of California at Berkeley. The comb structures are made of polysilicon parallel to the plane of the substrate. The structures are fabricated using the four-mask process but the masks are defined with a single mask. The process begins with a standard POCl₃ blanket n+ diffusion, which defines the substrate ground plane, after which the wafer is passivated with a 150 nm layer of low-pressure chemical vapour deposited (LPCVD) nitride on top of a 500 nm thick layer of thermal SiO₂ (a). Contact windows to the substrate ground plane are then opened (b) using a combination of plasma and wet etching. A 300 nm thick layer of *in situ* phosphorus-doped polysilicon is deposited by LPCVD at 650 °C then patterned with the second mask (c,d). A 2 μm thick LPCVD sacrificial phosphosilicate glass (PSG) layer is deposited and patterned with the third mask (e,f), which defines the anchors of the microstructures. The 2 μm thick polysilicon structural layer is then deposited by LPCVD (undoped) at 605 °C (g). The structural layer is doped by depositing another layer of 300 nm thick PSG (h) and then annealing at 950 °C for 1 h. A stress-annealing step is then optionally performed at 1050 °C for 30 min in N₂. The annealing temperature is lower than 1100 °C in order to avoid loss of adhesion between the PSG and the Si₃N₄. After stripping the top PSG layer by a timed etch in 10:1 HF, the plates beams, and electrostatic comb drive and sensor structures are defined in the final masking step (i). The structures are anisotropically patterned in CCl₄ plasma by reactive-ion etching, in order to achieve nearly vertical sidewalls. The wafer is immersed in 10:1 dilute HF to etch the sacrificial PSG (j). The wafer is rinsed repeatedly with distilled water (DI) at least 30 min after the micromachining step is completed and then dried in a standardized spin dryer.

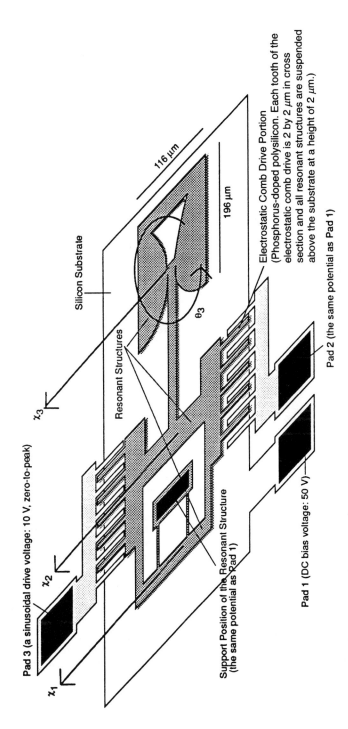

Fig. 3.6 Schematic drawing of a linear drive resonating structure with five degrees of freedom. Tests of the fabricated structures have demonstrated that peak-to-peak lateral displacements greater than 10 μm were feasible. The resonant frequencies of the structures varied from 1.7 to 33 kHz. (Modified from Brennen *et al.* (1990).)

Pad 3 (a sinusoidal drive voltage: 10 V, zero-to-peak)

x_3

x_2

x_1

Silicon Substrate

116 μm

196 μm

θ_3

Resonant Structures

Electrostatic Comb Drive Portion
(Phosphorus-doped polysilicon. Each tooth of the electrostatic comb drive is 2 by 2 μm in cross section and all resonant structures are suspended above the substrate at a height of 2 μm.)

Pad 2 (the same potential as Pad 1)

Support Position of the Resonant Structure
(the same potential as Pad 1)

Pad 1 (DC bias voltage: 50 V)

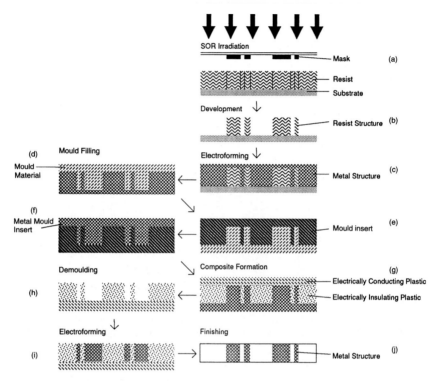

Fig. 3.7 Schematic representation of process steps of the LIGA (LIthographie, Galvanoformung, Abformung = lithography, electroforming, and moulding) method for mass fabrication of microdevices (adapted from Menz *et al.* (1991)). The LIGA process begins by generating a photoresist pattern by short-wavelength X-ray (wavelength = 1–10 Å) lithography on a conductive substrate (a). A heavy metal is used for the mask and some kind of PMMA (polymethyl methacrylate) is used for the resist. After removing the exposed resist, the gaps between the resist patterns can be preferentially electroplated right to the brim, yielding a highly accurate negative replica of the original resist pattern (b). In the next step, the resist structure is used as a template in an electroforming process where metal is deposited onto the elecrically conductive substrate (in the case of actuators, magnetic) (c–j). It is also possible to use the metal mould structure as a mould for such low-viscosity polymers as polyimide, polymethyl methacrylate or other plastic resins, or even ceramic slurries. After curing, the mould is removed, leaving behind microreplicas of the original pattern. A host of different materials are compatible with this process, and polymers, which are insulators, as well as metals may be used (Ehrfeld *et al.* 1988). The bottle-neck is the need for a short-wavelength, highly collimated X-ray source, typically a synchrotron orbital radiation (SOR) instrument. An extension of LIGA, the SLIGA (sacrificial layered LIGA) technique, gains another degree of design freedom by combining LIGA with the use of sacrificial layers. It is useful for making small gears, linkages, or other released parts that can be assembled on a separate LIGA structure or used in more traditional products.

'LIGA' stands for *Lithography, Galvanoformung, Abformung* ('lithography', 'electroforming', 'moulding').

If we use a short-wavelength electromagnetic wave in lithography we can produce a very fine structure in one resist. Applying focused X-rays on to some polymer instead of the conventional resist resin, we can obtain a new fabrication technique. The process was realized using X-rays of wavelength 0.2 to 2 nm from synchrotron orbital radiation as the etchant (Heuberger and Betz 1983; Wilson 1985; Heuberger 1986) and polymethyl-methacrylate (PMMA) as the X-ray resist (Ouano 1978; Sotobayashi *et al.* 1982; Emoto *et al.* 1985).

In Germany, Heuberger and Betz used a synchrotron (BESSY) energized to 0.8 GeV. They then developed a compact synchrotron for microfabrication (COSY), energized to 0.63 GeV. COSY had a 3 × 6 m synchrotron ring, and it was estimated that at 1 GeV COSY produced X-rays of wavelength 1 Å(λ_{max}) and 3.2 W mrad^{-1} and could fabricate structures 1 mm deep.

Ehrfeld *et al.* (1986, 1987) irradiated PMMA (0.5–3 mm thick, GS 233 or GS 279, Fa. Röhm) with X-rays through a gilded gold mask, which was cast in an X-ray absorbing ion-etch resist mask. Microparts were made by electroplating in the pattern after the irradiated PMMA evaporated. Nickel sulphate solution was used for electroplating (Fig. 3.7) and a honeycomb structure of depth 330 μm and width 80 μm was made. Becker *et al.* (1982) at the Institute für Kernverfahrenstechnik des Kernforshungszentrums, Karlsruhe developed a uranium mass-separation nozzle of depth 300 μm. ZFA Siemens also use this technique for microfabrication.

Recent work on the LIGA process has shown that combination with the silicon process is possible; this opens up the potential of novel applications for the fabrication of microturbines (Menz *et al.* 1991), gear trains, and three-dimensionally assembled structures used for filters, fluid-logics, and conduits. LIGA techniques are spreading gradually around the world (Guckel *et al.* 1991) and Microparts AG at Karlsruhe have begun to produce custom-made microparts.

3.3 Precision machining

Conventional precision machining is necessary to make fine parts using metals, alloys, ceramics, and bulk silicon.

Fine holes, gears, and turbines of micrometer size have been processed by microelectric discharge machining (MEDM) and wire electrodischarge grinding (WEDG) (Masaki *et al.* 1990) (Fig. 3.8). For example, electrode workpieces with complex shapes are possible and holes can be bored to a depth of about one-tenth of a wafer. In one case, WEDG was used to make a fine electrode 4.3 μm in diameter and 50 μm in length. Using this electrode, holes of 5 μm diameter were bored in a 20 μm thick stainless steel sheet using electric discharge processing (Fig. 3.9). The two-axis machining system was modified to a three-axis numeric

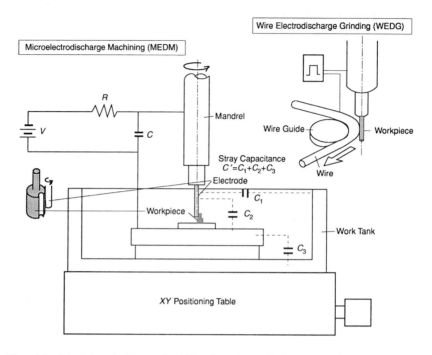

Fig. 3.8 Principle of three-axis NC microelectrodischarge machining and wire electrodischarge grinding. The discharge energy in a microdomain must be reduced to as small as 10^{-7}. As a result, the capacitance (C') should be reduced to 10 pF. To reduce the stray capacitance, the machine has been designed to minimize the use of metallic mechanical components. (Adapted from Masaki *et al.* (1990).)

control microelectrodischarge system by the Matsushita Research Institute Inc., Tokyo, and a microair turbine was fabricated and inserted into a metal catheter of external diameter 2.2 µm and rotated at 1000 rpm.

3.4   Laser-assisted etching

Laser light or an electron beam can be used for micromachining. These techniques are non-contact fabrication methods. If we make a scanner which can scan these beams in three dimensions, we can fabricate complicated three-dimensional microparts. As the precision of fabrication is mainly decided by the wavelength of the beam, beams with a shorter wavelength have been developed for photolithography or electron beam lithography.

Laser-assisted etching (LAE) is also a useful micromachining method. LAE utilizes a maskless photoreactive or thermal reactive process. The etching of

Fig. 3.9   A pyramid built on a steel ball using the microelectricdischarge fabrication method (courtesy Matsushita Giken Kogyo).

ceramics, semiconducting materials, metals, and high polymers in an etchant is enhanced using a laser beam. Laser-assisted etching does not use patterning masks. However, using conventional laser fabrication, the precision of fabrication is limited by the strong evaporation or melting caused by the high temperatures involved. Therefore, machines for LAE are commonly composed of a laser and an etching instrument. The fundamental processes are as follows:

1. laser irradiation in a liquid with etchant or in a gas gives rise to reactive ions or plasma;
2. the reactive ions or the plasma combine with the material or with the reactive atmosphere surrounding the material and change the material into a soluble or vaporized form.

The laser output power applied in this method is commonly very low, because the laser is used for only activating an opto- or thermochemical reaction and not for abrasion or burning off at high temperature.

The practical applications of laser-assisted etching as follows (Kobayashi 1988):

1. Ceramics fabrication: ceramics are easily fabricated using LAE. A ceramic can be etched at a rate of 200 μm s^{-1} in KOH solution using Nd–YAG laser

light. The method is applied to the fabrication of superconductive ceramic or diamond membranes (see Table 3.1 (a)).

2. Fabrication of semiconducting devices; using LAE, we can etch semiconducting devices 10 to 100 times faster than with usual etching procedures. As a 200 μm thick silicon wafer can easily be 'holed' using LAE, LAE is often applied to wafer fabrication. Some experiments have reported that anisotropic etching is possible by LAE (see Table 3.1 (b)).

3. Metal fabrication: an argon ion laser and wet etching are usually combined. As some metals are etched in etchant, the LAE etching speed is moderated in neutral salt solution or in atmospheric oxygen (Table 3.1 (c)).

4. Polymer fabrication: oxidizing fabrication using excimer lasers in atmospheric oxygen is frequently performed. Fabrication speeds of 1 μm per laser pulse are obtained for polyethylene terephthalate, polyimide, and other photoresists.

3.5   Photoforming and stereolithography

Photoforming is a universal optical forming technique and includes plain lithography and stereolithography. Photomask layering is a typical plain lithography technique and is frequently used in metal etching and silicon planar

Table 3.1(a)   Laser-assisted etching for fabrication of ceramics (adapted from Kobayashi (1988))

Materials	Laser	Etchant	Etching speed
Al_2O_3/TiC	Nd: YAG	KOH	200 μm/s
	Ar^+	KOH	200 μm/s
Ferrite	Nd: YAG	KOH	groove of 50 μm-w and 200 μm-d
	Ar^+	H_3PO_4	a hole of 100 μm-ϕ and 500 μm-d
PbTiZrO3(PZT)	KrF	KOH	same as Al_2O_3/TiC
	Ar^+, Kr^+	H_2	250 μm/s
Y–Ba–Cu–O superconductive materials	KrF	in air	0.1 μm/pulse
Diamond · hard carbon membrane	ArF	$Cl_2/O_2/NO_2$	0.2 μm/pulse
SiO_2	Ar^+	Cl_2	0.02 μm/min

Table 3.1(b) Laser-assisted etching for fabrication of semi-conducting devices (adapted from Kobayashi (1988))

Materials	Laser	Etchant	Etching speed
Si	CO_2	SF_6	experimental
		XeF_2	0.12 nm/pulse
	Ar^+	KOH	15 μm/s
		$Cl_2 \bullet HCl$	6 μm/s
		CF_4/O_2 (plasma)	
poli-Si	ArF	CF_3 Br $\bullet$ NFe	0.06 nm/puls
GaAs	Ar^+	HNO_3	30 μm/s
		CCl_4	6.3 μm/s
		H_2SO_4/H_2O_2	0.06 μm/min
	Nd: YAG	$CC1_4/H_2$ (RIE)	8 μm/min
	Ar^+	H_2SO_4/H_2O_2	8–18 μm/min
		$HNO_3 \bullet KOH$	
	ArF	CF_3 Br	2 μm/min
	HBr		8 μm/min
InP	H_3PO_4		groove 15×15 μm

Table 3.1(c) Laser-assisted etching for fabrication of metals (adapted from Kobayashi (1988))

Materials	Laser	Etchant	Etching speed
Fe	Ar^+	Neutral salt solution	1–3 μm/s
SUS			1.3–2.7 μm/s
SUS, Ni, Cu		HCL/HNO_3 solution	10^2–10^4 compare to ordinal melting
Al		$K_2Cr_2O_3/HNO_3/HPO_4$ solution	
W		in air	11.5 μm/s

process. The method is simple, but can easily be used to fabricate three-dimensional structure with submillimeter precision.

The technique can be applied to modify the characteristics of a polymer surface. Matsuda *et al.* (1989) at the Cardiovascular Center, Suita, Japan, used a UV photoreaction between phenylazide and nytren which can form covalent bonds to a C–C chain. After coating with fibronectin, collagen, or albumin, which have phenylazide chains, the material can be attached to carbohydrate using UV

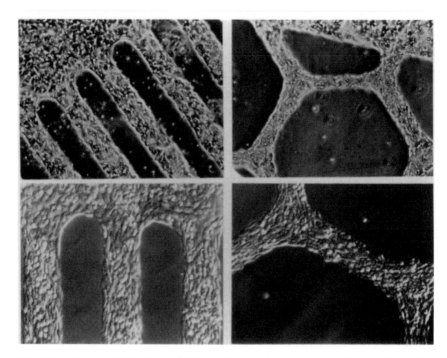

Fig. 3.10 Method of fabrication of fine cellular networks on a polymer plate using ultraviolet photolithography. This method is applied to the photochemical fixation of photoreactive polymers such as PDMAA (poly *N,N*-dimethyl acrylamide) (Matsuda *et al.* 1989).

radiation (250–320 nm) (see Fig. 3.10). Thus, the surface characteristics are modified by micropatterning when we use a photomask or a microoptical fiber for irradiation. As the fibronectin- and collagen-covered surfaces are hydrophobic and the albumin-coated surface is hydrophilic we can control the surface characteristics in the micro domain; this is useful for fabricating (constructing structure partially cells or tissues)

Ultraviolet laser beams scanned onto a thin layer of photopolymerizing solution can produce two-dimensional structures. After hardening, more polymer solution is added to the initial solution and the surface is UV irradiated with another pattern, resulting in a single three-dimensional structure. This method is called stereolithography and was first developed by 3D Systems Inc., Santa Clara, California. X-rays of wavelength 307 nm emitted from an excimer laser were applied on epoxy or urethane resin diluted and mixed with a benzoin or acetophenone photoreactor (Fig. 3.11). The method is called the free-surface method and the minimum solution thickness is reported to be 200 μm. In order to

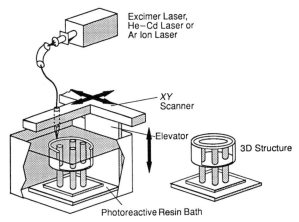

Fig. 3.11   A macrophotoforming technique. The method is a prototype of automatic three-dimensional stereolithography. A photohardening polymer liquid such as phenylazide is stored in a bath and the thin layer is hardened with ultraviolet radiation which is scanned according to the tomographic images of a computer model. (Adapted from Nagamori (1990).)

produce microstructures, the fixed surface method was developed (Takagi and Nakajima 1993; Ikuta and Hirowatari 1993). The apparatus for this process consists of a UV laser light (He–Cd laser, CW, 325 nm wavelength, 10 mW output) or a xenon lamp X–Y scanner and a resin container in which a computer-controlled platform is installed. The laser light penetrates the bottom of the container or an upper window, coated with PFA resin to avoid resin adhesion, and the model is fabricated from top to bottom or bottom to top (Fig. 3.12). Using the method, a minimum box size of less than 10 μm^3 is possible, with 0.3 mJ per box input. Cords, bending pipes, microcoil springs, combs, microvalves, a see-saw, a turbine, and a helicopter of 2 mm length have been made using this method.

3.6   Manufacturing and handling under microscopes

3.6.1 Microdrip fabrication

Eumenes micado is a kind of wasp which makes a nest of soil in the shape of a narrow-necked bottle. This is an example of 'microdrip' fabrication, which can now be reproduced by some engineers and handicraft workers. For micromachine fabrication, Sonin and co-workers of MIT have started a droplet-based manufacturing research project. They modified an ink-jet printer to make three-dimensional models. The wax is heated to around 90 °C to melt it and is ejected

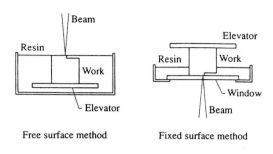

Free surface method   Fixed surface method

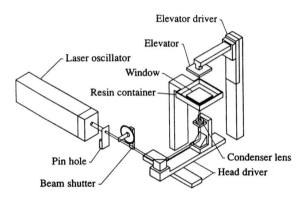

Fig. 3.12 Free surface and fixed surface methods of micro photoforming. Prototype models (comb, see-saw, turbine, helicopter) are photoformed by an He–Cd laser (continuous radiation, 325 nm wavelength, 10 mW output) to acryl polymer at a maximum rate of 40 cells per second and with 1 μm step scanning (Takagi and Nakajima 1993).

through the ink-jet head in droplets around 50 μm in diameter. With these droplets they made 300 μm high wax letters, 'MIT' (Fig. 3.13). Melissa Orme and a group at the University of Southern California (USC) have built their own sophisticated droplet generator that can melt and spray materials like Rose's metal, an alloy of bismuth, lead, and tin, that has a melting point of around 190 °C. With it, they demonstrated the ability to make microscopic metal pipe fittings. MIT engineers have also been working with microdroplets of a tin alloy and have already built the next generation of droplet makers. These machines will be able to discharge droplets of copper, a metal that melts at around 1100 °C. Once the technology has been tested with high melting-point metals, industry will take a closer look at it; an

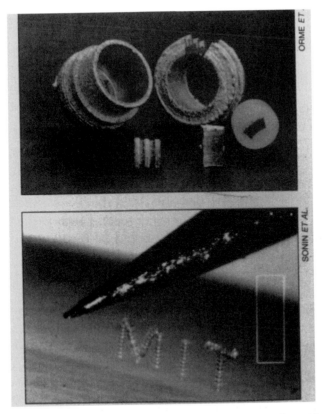

Fig. 3.13 Microdrop forged metal fittings and wax letters. (Photos adapted from Travis (1992).)

engineer might someday be able to design a part on a computer display, then 'print' it out as a finished object, much as office workers now turn out memos on an ink-jet printer (Travis 1992).

3.6.2 Manufacturing using scanning probe microscopes and electron microscopes

Hatamura and co-workers at the University of Tokyo have developed a fabrication system for three-dimensional microstructures using a scanning electron microscope (SEM). They developed a stereoscopic observation instrument modified from an SEM and used the vacuum chamber of the SEM as a working table. On the table they built a lithographic etching instrument called a multi-face fast atom beam (FAB), a micromachining centre containing robot hands with four rotational and three translational degrees of freedom, and many mechanical tools such as diggers, tweezers, blow pipes, scrapers, and sticking tools. They also made a micro five-storey tower, 280 μm high and 100 μm wide and deep, shaped like the tower of a Japanese temple, using a GaAs wafer etched by the chlorine ion beam of the multi-face FAB. The FAB mask was an etched nickel plate 10 μm thick. The etching speed is 0.15 μm min^{-1} at 60 °C. The workpiece was fixed on a rotating Invar alloy holder, and etched. It was rotated by 90° and etched once more. The tower was cut off by some microtools, moved by robot hands, and installed in a support hole 50 μm square excavated on a polyimide substrate using a vibrating digging tool and UV laser irradiation from a Kr–F excimer laser. They are planning to develop a new concept of the factory, a 'micro factory', applying such instruments, and to produce a manufacturing system similar to the system used in a conventional mechanical factory (Hatamura *et al.* 1994; Fig. 3.14).

3.6.3 Handling of microparticles with laser tweezers

Absorption, reflection, or refraction of light by a dielectric material generates tiny forces. Several milliwatts of power produced by a strong laser light generates a force of only a few piconewtons (1 pN = 0.1 μdyn). However, a force of 10 pN is sufficient to pull a microsphere through water ten times faster than *Escherichia coli* swim. So in the microscopic domain, radiation pressure could play a significant role in the handling of microscopic particles (Block 1992).

In 1980, Ashkin and colleagues at AT&T Bell Laboratories proposed a laser-based optical trap for microscopic particles. They demonstrated that it could non-invasively manipulate living material, including viruses, bacteria, yeast, and protozoa. This invention makes use of the so-called 'gradient force' that appears in a light gradient when a transparent material with a refractive index greater than the surrounding medium is placed in it. This principle has been applied in optical tweezers, micromanipulators, tensiometers, etc.

As light passes through polarizable material such as dielectrics, it induces fluctuating dipoles. These dipoles interact with the electromagnetic field gradient and produce a force which directs microparticles towards brighter regions. When we direct parallel laser light onto a microsphere from above, the light is bent because the sphere acts as a lens. If the intensity profile of the incoming beam is uniform, the reaction forces on the left- and right-hand sides of the sphere cancel and there is no net sideways component. However, asymmetry in the light field gradient causes an imbalance in the reaction forces and the object is pulled towards the bright side. Such forces have been used in optical tweezers. In order to move the sphere against the light-scattering force sharply focused light is required. Such a sharp focus can be achieved by focusing laser through a microscope objective of high numerical aperture light on to a diffraction-limited spot having a diameter of about the wavelength of the light. Diode lasers in the near-infrared region (780–950 nm) makes these devices practically attractive for low-power use. The instrument can be combined with any number of microscope imaging modes, including bright field, phase contrast, and epifluorescence. Laser tweezers are relatively free of creep, backlash, and hysteresis and they remain confocal with the specimen.

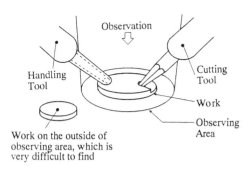

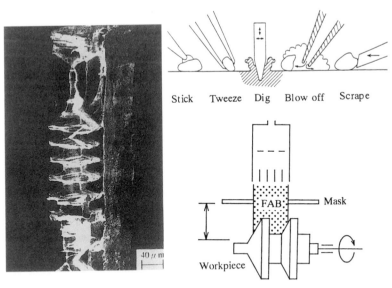

Fig. 3.14 Three-dimensional microstructures fabricated by a multiface fast atom beam (FAB) on a stereoscopic scanning microscope stage. Various mechanical tools, a three-dimensional microassembly working table, and the three-dimensional shape-making method by multiface FAB were used for production of micro five-storey towers of a Japanese temple. (Figures and photograph adapted from Hatamura *et al.* (1994).)

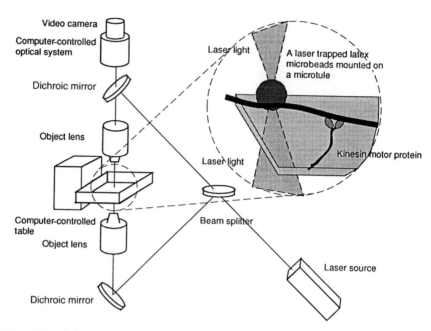

Fig. 3.15 A laser tweezer being used to handle an organella and measure the dynamics of a motor protein.

Today, laser tweezers are used in the assembly of microparticles and to measure the dynamics of movement of micromechanical particles, organelle motion, and forces produced by motor protein (Fig. 3.15).

4
Machines incorporating micromechanical devices

4.1 Microsensors

Initial research work on micromechanical devices led to the development of microsensors using the silicon process. Muller (1987) report sensors with movable components, such as cantilevers, diaphragms, and microbeams, proof masses (one kind of weights) and so forth which were developed at the University of California, Berkeley (UCB). The sensors were fabricated on silicon wafers using anisotropic etching in hydrofluoric acid (see Fig. 3.5). Thin film diaphragms made by the silicon planar process are frequently used in pressure sensors and tonometers (Fig. 4.1).

Silicon accelerometers with a silicon proof mass are used in automotive applications and many other mobile instruments (Esashi 1994). The bulk-processed microproof mass is suspended by silicon beams, and piezoresistors on the

Fig. 4.1 Cross section of a microfabricated silicon diaphragm pressure sensor. The membrane (Si$_3$N$_4$ diaphragm) was 80 × 80 μm in width and length and 1.6 μm thick. (Photograph adapted from Igarashi (1990).)

suspension beams sense the motion of the proof mass caused when acceleration stops or changes (Roylance and Angell 1979). Piezoresistive accelerometers are cheap and have a wide frequency range but are of low precision; they are the basis for anti-skid and air-bag systems. Capacitive accelerometers have a high sensitivity but a small dynamic range, because the gap between the capacitor plates is made as small as possible (Suzuki *et al.* 1990). If we wish to extend the dynamic range, we must fix the mass at the initial position using a feedback force. Esashi and co-workers developed a multiaxis capacitive accelerometer with a surrounding seismic mass and used it to detect acceleration in three directions at once (Jono *et al.* 1994).

The principle of sensing a resonant frequency with a resonating proof mass has also been applied to detection of gases and infrared radiation (Cabuz *et al.* 1993). A polysilicon bridge coated with a high polymer is made to oscillate and the mass of materials absorbed in the polymer is detected by the phaseshift of resonating beams caused by a change in capacitance (see Fig. 3.6). The IR sensor consists of two thin silicon dioxide beam bridges on a cold silicon rim; the resonant frequency of the bridge changes due to strain variation caused by IR radiation.

Mechanical sensors are essential tools for developing and evaluating micromachines. Micromotor torque, the stretch force of a linear microactuator, pressure in a microconduit, and so on must be measured. Microtorque is a good parameter to measure in a micromachine. Mikuriya *et al.* (1993) reported a microtorque measurement system. They used a force meter coupled to a meter shaft floated on air bearings to reduce friction and measured torque over a working range from 100 nN m to 10 mN m with a resolution of 10 nN m. Gass *et al.*, at the University of Neuchâtel, Switzerland, developed a microtorque sensor based on differential force measurement, which can measure forces with resolution of 0.05 µN m over a working range from –200 µN m to 200 µN m (Gass *et al.* 1994). First they made a microforce sensor to measure microflow (Gass *et al.* 1993). The force sensors were silicon cantilevers fabricated by standard planar technology and bulk etching technology, with resistors placed in the bending suspension of the cantilevers. The sensor consists of two resistive force sensors. A spring blade made of copper–beryllium alloy having a thickness of about 100 µm mounted perpendicular to the torque axis converts the torque to a force acting on the two force sensors. The resistors are connected in a Wheatstone bridge system and the voltage drop over the bridge is amplified. This device has been coupled to a hybrid integrated micromotor and the dynamic torque was measured to be 1.4 µN m (Rachine *et al.* 1993).

4.2 Microfittings

Microsize pin joints, sliders, gears, and springs were the next micromechanical devices to be developed. The fabrication techniques used for microsensors were applied to micromechanical systems. A set of simple pin joints and cranks made by

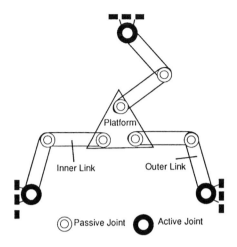

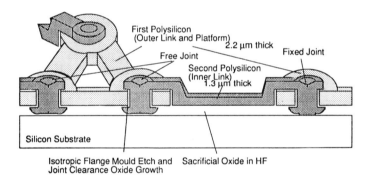

Fig. 4.2   The mechanism and fabrication process of a paralled mechanism with three degrees of freedom. The fabrication protocol is as follows. Initially, 2.2 μm of low-temperature SiO_2 (LTO) is deposited followed by the first LPCVD polysilicon deposition. The polysilicon is heavily doped with phosphorus and patterned in RIE (reactive ion etching) to form the outer links and the output platform. A thin, thermally grown oxide etch mask is used for the polysilicon RIE etch step. Where the pin joints and the bearings are to be formed, the LTO is isotropically etched to a depth of 1 μm. The polysilicon layer is thermally oxidized a second time to provide 0.3 μm oxide coverage in the pin joints, corresponding to the pin-joint clearances in the final device. A second LPCVD polysilicon layer is deposited, heavily doped with phosphorus, and patterned to define the inner links and pin joints. Finally, the mechanism is released by dissolving the sacrificial oxide in HF. When released, the weight of the mechanism is supported by the flanges in the three fixed joints and the undersides of the six free joints. (Modified from Behi *et al.* (1990).)

Muller at UCB was the first of many micromechanical systems. A pin joint 25 μm in diameter and cranks were made of polysilicon fabricated by low-pressure chemical vapour deposition and sacrificial layer etching of phosphosilicate glass (Fan *et al.* 1988). Using this process they developed a 100 μm slider, a combination of gears and slider, and a crank with a slot and a spring, and built up a microlink structure with three degrees of freedom (Behi *et al.* 1990) (Fig. 4.2).

Another early micromechanical device to be developed was the set of gears developed by Trimmer and co-workers. Trimmer, who guided teams at the Robotics Research Department of AT&T Bell Laborotories and the Massachusetts Institute of Technology, made three gears of 125–240 μm diameter (Meheregany *et al.* 1988). The same techniques were used for electrostatic micromotor production. As shown in Fig. 3.4, they first deposited a layer of polysilicon on a silicon substrate coated with silicon nitride for the anchor of the gear; a layer of phosphosilicate glass was then deposited. Anchor areas were then etched in two mask steps. A 4.5 μm thick polysilicon layer was then deposited and patterned to form the gear and wall. To create clearance for the bearing, another sacrificial layer was deposited and patterned. A third polysilicon layer formed the bearing. The glass was etched away with hydrofluoric acid in order to release the gear (Tai *et al.* 1989).

The LIGA process could expand the field of batch-fabricated micromechanical devices to truly three-dimensional systems. Guckel *et al.*, of the Wisconsin Center for Applied Microelectronics, University of Wisconsin, Madison, developed align-and-clamping jigs for maintaining the alignment of an X-ray mask to the optically defined sacrificial pattern. They employed very thick PMMA photoresist layers, up to 500 μm, and made free gears, fully attached shafts, and many nickel microstructures. They assembled more complex structures using the fittings (see Fig. 3.7).

4.3 Microactuators

One of the most important components of micromachines is the microactuator. A basic problem in dealing with microdynamic systems is the scaling of physical properties in the microscopic realm. A mechanical engineer's experience of what is and is not possible has been formed by observing how things work in the macro world. Is this experience applicable in the micro world? Let us consider a dynamic actuator that has been miniaturized by a factor of 10^3. The inertial and gravitational forces, which are proportional to the mass, decrease by a factor of 10^9. Forces that scale as the area of the system, such as surface tension and electrostatic attraction, will scale down by a factor of only 10^6. The ratio of the latter forces to inertial forces increases by a factor of 10^3 (see Fig. 2.2). For example surface tension prevents water in a microcup from escaping and micro particles on the surface of a micromirror are held there by an electrostatic force that is much stronger than the gravitational force on them. When we design a microsystem, we must find forces useful for actuation taking into account such scaling laws.

4.3.1 Electromagnetic microactuators

In the macro world, electromagnetic forces are dominant and are applied in conventional motors. But electromagnetic motors utilize a constant current density in the motor windings, and this scales in proportion to the area of the winding squared. The power shrinks markedly as such motors become smaller. As a result, models of normal electromagnetic motors are inadequate for application in micromotor design. But size reduction greatly reduces heat dissipation, which scales with the area, so it may be possible to increase the current density with this system. Much work towards realizing practical micromotors has focused on electrostatic drive, but recently there have been some investigations using different principles for a variety of applications. Work has been published on ultrasonic, dielectric induction, and electromagnetic drive in MEMS proceedings.

Electromagnetic actuators are attractive for operation in environments where high driving voltages are unacceptable or unattainable, such as in conductive fluids for some biomedical applications. But technically, it is difficult to fabricate the necessary three-dimensional wrapped coils using an integrated, planar fabrication process.

Wagner and co-workers, at the Fraunhofer-Institut, Berlin, made two chains of planar coils on a silicon wafer by a local electroplating method. A sliding magnetic block, $0.7 \times 1.8 \times 0.9$ mm^3, or a rolling magnet with a diameter of 1.0 mm and a thickness of 0.75 mm, was placed between the coil chains. During rotational magnet testing, two glasses fabricated planar magnet chains sandwiched the magnet in the gap of a width of 0.75 mm and rolling a velocity of 2.5 cm/s. They also tried to arrange planar coils in a ring or rectangle and to roll a permanent magnet in the ring. The permanent magnet was about 0.3 mm in diameter and the overall size of the actuators was less than 1 mm. They observed synchronous rotation up to 2000 rpm for a current of 500 mA (Wagner *et al.* 1992).

Instead of planar coils, a meander-type inductive component has been made using an integrated planar process (Ahn *et al.* 1993). Ahn *et al.* made 12 stator poles with meander-type induction coils and 10 rotor poles. The stator and pin were fabricated using a polyimide multilevel metal interconnection technique, in which an electroplated high-permeability Ni(81%)–Fe(19%) magnet core was used as the magnetic material. A 40 μm thick rotor 600 μm in diameter rotated at speeds of up to 500 rpm with applied currents of 300–500 mA and a driving voltage of less than 1 V. The predicted torque was estimated as 1.2 μN m for a stator current of 500 mA (Fig. 4.3).

Recently, a practical planar rotational magnetic micromotor has been developed by Guckel *et al.* at the University of Wisconsin (Guckel *et al.* 1993). They made two magnetic motors with six stator poles and four rotor poles. The rotors were 285 μm and 423 μm in diameter and the stator and rotor heights were up to 300 μm. The motor was constructed using a deep X-ray lithography and electroplating process or LIGA and was integrated with standard IC processing

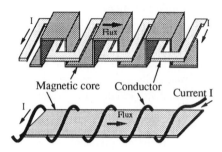

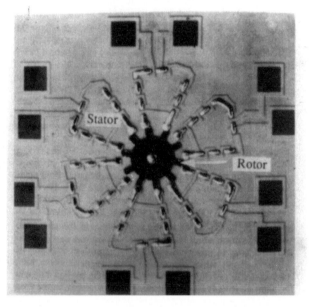

Fig. 4.3 Integrated assembly of a magnetic micromotor requires the innovation of micro-electromagnetic coils. A meander-type inductive component is fabricated using a polyimide thick resist layer which is photolithographed by laser-assisted etching. In the production technique, the resist acts as a mould and the actual layer is fabricated by electroplating. A multilayered magnetic micromotor is easily produced using this technique. (Figures adapted from Ahn *et al.* (1993).)

(Christenson *et al.* 1991). The stator and rotor were made of pure nickel and fabricated by the sacrificial LIGA process. The stator was installed between electroplated Ni poles, and wires were bonded between the poles for making the induction coil structure. The motor, with a rotor diameter of 423 µm, operated at a maximum speed of 12 000 rpm and the 285 µm rotor rotated at speeds slightly above 30 000 rpm, both at current excitation levels of 0.6 A (peak). The rotor showed no evidence of change in operation after 50 million rotations (Fig. 4.4).

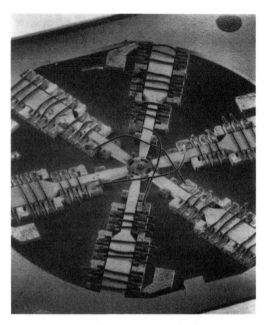

Fig. 4.4   An electromagnetic micromotor produced by the sacrificial LIGA process (from Christenson *et al.* (1991).)

4.3.2 Electrostatic microactuators

Electrostatic attractive forces scale with area. Moreover, breakdown of the electric field has been shown to increase in microgaps by a factor of more than ten times the macroscopic limit, which is usually 3 MV m^{-1}. The result is an even more attractive scaling for electrostatics. As this scaling analysis suggests, electrostatic forces are more suitable than electromagnetic forces in the micro world. Electrostatic micromotors of various designs have been developed within this framework.

Since 1987, when the first silicon electrostatic micromotor was designed by Lang at the Laboratory for Electromagnetic and Electronics Systems of the UCB, several silicon micromotor research efforts have been undertaken. The first variable-capacitance micromotor system was reported at UCB. This system has a four-poles rotor which spins very slowly around the central hub or bearing. The motor has a 120 μm diameter rotor which is 2 μm thick. The motor is made of polycrystalline silicon, while phosphosilicate glass serves as the sacrificial layer, etched away after all the layers have been deposited (Tai *et al.* 1989). Recently, at MIT, a 100 μm diameter eight rotor has been rotated at 2500 rpm in response to varying voltages (the starting voltage was 74 V) that were applied sequentially to the twelve stator poles encircling it, across a gap of 2 μm (Fig. 4.5) (Meheregany *et al.* 1990).

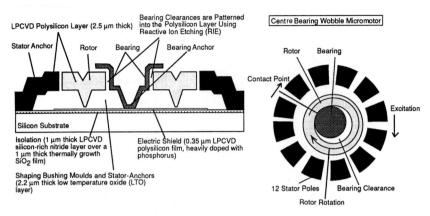

Fig. 4.5 A micromachined harmonic side-drive motor. This figure shows a typical micromotor cross section summarizing the fabrication process. Initially, substrate isolation is established using a 1 μm thick LPCVD silicon-rich nitride layer over a 1 μm thick thermally grown SiO_2 film. A thin (350 nm) LPCVD polysilicon film is deposited, heavily doped with phosphorus, and patterned to form the electric shield. A 2.2 μm thick low-temperature oxide layer (LTO) is deposited and patterned in two steps to form the stator anchors and the bushing moulds. A 2.5 μm thick LPCVD polysilicon layer is deposited and heavily doped with phosphorus. The rotor, stator, and air gaps are patterned into this polysilicon layer using RIE. The thickness of the polysilicon rotor–stator layer is 2.2 μm at this point, since a patterned thermally grown oxide mask is used for the RIE etch of the polysilicon. The second sacrificial LTO layer is deposited, providing an estimated 0.3 μm coverage on the rotor inside radius sidewalls, and patterned to open the bearing anchor. A 1 μm thick LPCVD polysilicon film is deposited, heavily doped with phosphorus, and patterned to form the bearing. Finally, the sacrificial layers of LTO are released in HF. The wobble motor with 12 stator poles and 8 rotor poles, a 1.5 μm wide air gap, and 130 μm diameter is driven by a six-phase, unipolar, square-wave exitation. The exitation voltage as high as 150 V across the air gaps (i.e. electric field intensities of 1×10^8 V m^{-1}) are used without electric field breakdown in the air gaps. The 12:8 side-drive motors are spun periodically with signal frequencies up 167 Hz corresponding to a rotor speed of 2500 rpm. The motor occasionally spun 15 000 rpm at 1 kHz exitation. However, the motor typically operated up to 140 rpm. (Adapted from Meheregany *et al.* (1990).)

A linear electrostatic micromotor was first developed by Fujita at the University of Tokyo (Fig. 4.6). A silicon substrate slider was fabricated with a permanent magnet strip. The slider had a 100 μm width glide on plane stator with a 100 μm strip of YBCO high-temperature superconductor, using a form of a magnetic levitation known as the Meissner effect (Kim *et al.* 1989).

If a flat stator is rolled up and a rotor inserted into it, varying static voltages can then be applied sequentially to the stator pole with the result that the rotor rolls over the stator. The rotor then turns harmonically. This so-called harmonic motor, also known as the wobble motor, is practically applicable to the microactuator today. Jacobsen of the University of Utah, Salt Lake City, has reported many variations and fabrication methods for the wobble motor (Jacobsen *et al.* 1989): a

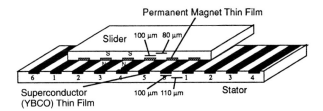

Permanent Magnet Thin Film

Slider   100 μm   80 μm

S   S

6   1   2   3   4   5   1   2   3   4

Superconductor   100 μm   110 μm   Stator
(YBCO) Thin Film

Fig. 4.6 Schematic diagram of a superconducting actuator using the Meissner effect. In the stator, YBCO superconductors of a few micrometers thick, 100 μm wide and 2 mm long are patterned by photolithography. When the stator is cooled by liquid nitrogen, the slider, which has permanent magnetic stripes levitated by the Meissner effect, can be driven by switching the on–off signals of the superconducting stripes. (Adapted from Kim *et al.* (1989).)

1250 μm diameter, 15 mm long rotor was inserted and spun at 250 turns per minute in a stator of 1350 μm inner diameter. The stator hole was surrounded by 32 copper wires 135 μm in diameter which were moulded in epoxy resin.

The concept of the electrostatic actuator expands the capacity of a multilayered electrostatic film actuator and was developed by Egawa and Higuchi at the University of Tokyo (Fig. 4.7). A set of multielectrodes at intervals of 0.1 mm was deposited on both sides of a stator film. Between the stator films polyethylene terephthalate (PET) film was inserted to act as a slider. When three successive electrode charges—positive, negative, and neutral electrostatic voltages respectively—are applied, a set of mirror images of negative, positive, and neutral voltages are induced on the opposite positions on the PET film. When a slightly conductive material such as PET is used the mirror images remain for a short time after the stimulated charges have been withdrawn. After a short interval, the charges on the electrode change to a new order—neutral, positive, and neutral—and the slider lifts up and moves toward the slider negative image. If the timing of the one-step movement is matched within a short period, the image negative charge comes to the position of slider negative charge and produces a repulsive force.

A dispersed and integrated actuator system has been developed by Esashi at Tohoku University, Sendai, Japan. The basic origin of the actuator driving force is the electrostatic attractive force of two curved electrodes facing each other. Esashi and co-workers tested a macromodel using curved polyimide films (Kapton®, Dupont), 25 or 50 μm thick, 80 mm long, and 30 mm wide, and on which 25 or 16 nm thick Ni electrodes were sputtered. Five Ni-sputtered polyimide films are stacked and stuck 10 mm apart. The device became 32 per cent thinner with a power supply of 500 V and produced a force of 4.5 mN (Yamaguchi *et al.* 1993; see Fig. 4.8). The same idea was reported as 'integrated force arrays' by Bobbio at the University of North Carolina, Charlotte, in 1993. The technology is in an early stage of development but large-scale integration of actuator cells is expected. One unit cell is made of two polyimde beams 0.5 mm thick, between 2 and 4 mm high,

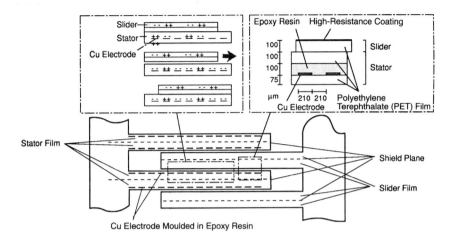

Fig. 4.7 Structure of a multilayered actuator and the principle of image charge stepping action. The actuator has a stator in the form of multilayered films which contain wire electrodes and multilayered slider films which do not contain wire electrodes and consist of isolator and high-resistance layers. The slider put on the stator has no electric charge at first. Positive and negative voltages are applied on two groups of of stator electrodes. The electric field produced by the slider's electrodes causes movement of charges in the resistor which continues until the induced charges neutralize the field in the resistor. In the equivalent state, the induced charges can be indicated by a mirror image of charges on the electrodes. After sufficient charges are induced, the signs of voltages on the electrodes are inverted from positive to negative or negative to positive. Although the electric charges on each electrode are instantly changed, the charges in the slider resistior cannot move in such a short time because the charges are obstructed by the high resistance of the isolator layer. Thus, mirror images of charges are produced between slider and stator and the slider levitates on the stator film. The motive force is produced in the same way as that in a linear stepping motor. This actuator has many merits for the development of microactuators. This actuator is easy to integrate, and does not need alignment between films, feedback control, nor high precision in the dimensions of the electrodes. (Adapted from Egawa and Higuchi (1990).)

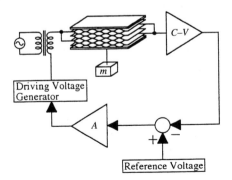

Fig. 4.8 An electrostatic microactuator piled multiple layers of polyimide film. (Adapted from Yamaguchi *et al.* (1993).)

and 7 to 21 mm long. Two polyimide beams are chained each other and make large polyimide cross bridge networks. On one side of a polyimide beam metal electrodes are sputtered or doped to make a capacitor. The model is estimated to deform by 30 per cent of its length and 1 cm^2 of its area; 1.5 million elements produce an average force equivalent to 0.4 G. This actuator is a good model of a VSLI actuator and might be used to compete with the working of animal muscle.

4.3.3 Piezoelectric actuators

The piezoelectric effect is also applicable in microdynamic systems. A microfabricated 1 mm cantilever beam of silicon nitride developed at Stanford University for scanning tunneling microscopy (STM) moves in three dimensions when voltages are applied to four independently addressable regions of piezoelectric zinc oxide (Quate 1990). The same idea has been applied to stereoscopic teleoperation in a scanning electron microscope (Hatamura and Morishita 1990) and to the micronization of an impact drive mechanism in a microinjector system (Higuchi *et al.* 1990).

Microfabricated valves were first reported in 1979 by Terry at Stanford University, Palo Alto, California in a silicon gas chromatography project. Recently other IC-based valves have been under intensive development by Esashi at Tohoku University (Fig. 4.9), and by Fluitman at the University of Twente, Enschede, The Netherlands (Elwenspoek *et al.* 1989). Tohoku researchers have demonstrated a variety of flow-control systems built around one-way normally closed valves with suspended polysilicon check rings and fabricated using a

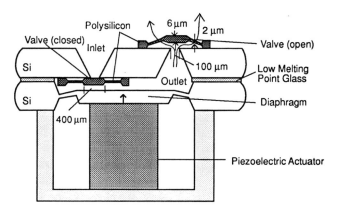

Fig. 4.9 Cross section of a micropump fabricated on silicon substrates The unit consists of two normally closed microvalves, a silicon fluidic channel, and a diaphragm driven by a piezoelectric actuator. The microvalve unit is made of Pyrex glass and silicon, and the n-type silicon wafers, 280 μm thick, are etched and formed into a fluidic device. The one-way valves are micromachined using polysilicon and PSG deposited by LPCVD. A valve unit can control gas flow between 0.1 ml min^{-1} and 85 ml min^{-1} at a pressure of 0.75 kgf cm^{-2} pressure. The piezoelectricdriven diaphragm pump can pump 20 μl min^{-1} against a pressure of 780 mmH$_2$O cm^{-2}. (Adapted from Esashi *et al.* (1989).)

combination of bulk and surface micromachining techniques. To control the diaphragm displacement, miniature piezoelectric actuators are mounted on the silicon micromachined wafer. A diaphragm-type pump consisting of two polysilicon one-way valves and a piezoelectric actuator was produced. The maximum pumping flow rate and pressure are 20 ml min^{-1} under a load of 780 mm H_2O cm^{-2}.

When two 2–3 cm arch-shaped sheets of piezoelectric polymer film were stuck to each other and 10 V AC applied between the sheets, the sheets began moving very fast. The difference in friction between the surfaces of the two sheets produced the movement. This phenomenon might be utilized in many mobile microstructures (Hayashi 1985).

4.3.4 Thermally actuated micromechanisms

A 0.5 mm layer of p$^+$ silicon substrate covered with four polysilicon electrodes was fabricated by the silicon process and actuated thermally as a bimetallic switch (Reithmuller and Benecke 1988). In this microstructure, thermal expansion produces a large power. Themal expansion can also provide the force for seating valves or displacing diaphragms in microfabricated flow-control actuators. For example, researchers at the University of Twente, expanding on earlier work at Stanford University, are using the thermal expansion of gas to actuate a pump (Elwenspoek *et al.* 1989).

Shape memory alloys (SMA) and shape memory materials have wide application in the fields of industry and medicine. Recently, an SMA has been deposited on a silicon wafer and fabricated in a fine wire, making it possible to use SMA for microactuators. Practically, SMA research was initiated in 1963 with Ti–Ni (Buehler *et al.* 1963), but rapid progress in development has taken place in the last three years (Ikuta *et al.* 1990). The main applications so far have been for artificial muscles in an artificial finger and in the bending of flexible endoscopes.

4.3.5 Photothermally actuated microactuators

Mizoguchi *et al.* reported a photothermally actuated microactuator in 1992. The actuator consisted of an SiO_2/NiCrSi diaphragm 0.7 μm thick covering a 250 μm deep cell processed by the silicon process (Fig. 4.10). The cell is filled with thermally phase-modulated material (Fulorinat FC72) and carbon fibre for absorbing light energy. The diaphragm measures 400 by 800 μm and can with stand a pressure of 70 kPa. The carbon fiber was irradiated with 40 mW fiber-guided laser light which vaporized the medium. Mizoguchi *et al.* developed a micropump consisting of five light-driven actuators. It can pump out 0.58 μl min^{-1} against a 10 mm H_2O head using a 3 Hz, 40 mW laser light input. Light-driven actuators could be applied in *in vivo* medical microequipment in order to prevent electric microshocks.

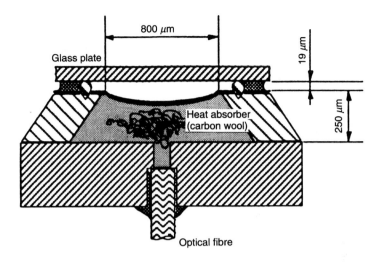

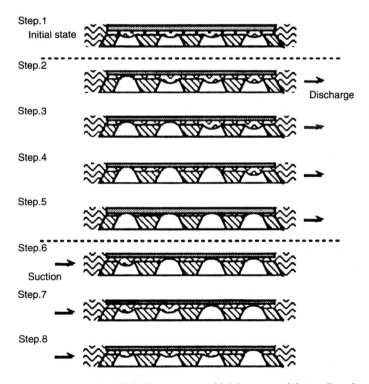

Fig. 4.10   A micromachined multiple linear pump which is actuated thermally using an input of laser light power. (From Mizoguchi *et al.* (1992).)

4.4 Microfluidic devices

There are many demands for integrated microfluidic devices, not only in analytical chemistry and microchemical production but also in cell handling and gene technology. Herzenberg *et al.* (1976) assembled a micronozzle, ultrasonic nozzle vibrator, laser beam, and microelectrode and developed fluorescence-activated cell sorting (FACS). This was a good example of micromachine assembly.

The integrated pneumatic or fluid circuit is a good example of early micromachine applications. Nozzles, conduits, connectors, distributors, injectors, valves, and so forth have been developed and applied to micropneumatic and fluid instruments such as gas and liquid chromatographies, cell sorters, and autoanalysers. A microfluidic amplifier was studied in the 1960s (Kirshner 1966) and applied in a pressure controller. Gas chromatography on a silicon wafer was reported by Terry *et al.* (1979).

4.4.1 Microvalves

Many integrated fluid devices and their parts have been developed since micromachine research began. Some microvalves are driven by electrostatics (Ohnstein *et al.* 1990; Huff *et al.* 1993), thermomechanics (Trah *et al.* 1993), piezoelectrics (Shoji *et al.* 1991, and Fig. 4.9), or magnetically (Yanagisawa *et al.* 1993) for microgas separation and fluid control systems. The magnetic and the piezoelectric driven valves require the hybrid integration of magnets (Wagner and Benecke 1991) or piezodiscs to achieve a large deflection.

Electrostatic excitation can easily be integrated by monolithic processes such as molecular beam epitaxy, low-pressure chemical vapour deposition, and reactive ion etching, which require precise control of ultra clean gas flows at very low flow rates, as do many semiconductor processes. However, the distance of movement of the valve is limited by the short-range force. As an example, a seven-channel microvalve was fabricated on a chip, one valve included a bridge valve of beam length 300–1400 μm. This valve controlled a gas flow of 1 SCCM at 100 Torr with leak rate of 0.006 SCCM. The valve clearance was only 12 μm and it was driven by 60 V DC (Robertson and Wise 1994).

More sophisticated electrostatic valves use S-shaped (Shikida *et al.* 1993) or pre-buckled electrodes (Branebjerg and Gravesen 1992). However, using thermal deflection we can easily get both high forces and large deflections in a simple way (Jerman 1991). Such a microvalve, containing an electrically actuated bimorph diaphragm, was made to provide a controlled gas flow up to 150 ml min^{-1} in a pressure range up to 3.5 bar with an input of less than 0.5 W. The drawback is that the response time of this type of valve is too long for switching applications. Improving such defects, Lisec *et al.* (1994) developed a thermal driving principle using the buckling effect of a silicon microbridge. The in microvalve consisted of two $5 \times 5 \times 0.5$ mm^3 silicon chips. The valve membrane structure had an area of

$2600 \times 2600 \ \mu m^2$ with n^+ doped polysilicon heaters and gold metal lines, and shuts the valve seat with an orifice of $360 \times 360 \ \mu m^2$. The valve was fabricated from a 525 μm thick polished silicon wafer and required seven photolithography masks. The valve beam thickness was around 12 mm and four 600 nm polysilicon heaters were patterned on an oxidized substrate. The response time was estimated to be about 15 ms and measured 3.3 Hz for a pressure of 0.5 bar.

4.5 Energy supply and sources

When micromachines are developed we must consider their energy supply methods and energy sources. In practice we can use electric energy supplied through wires or electromagnetic wave energy through cables. However, when we want to develop stand-alone micromachines, we must use batteries or energy acquisition systems using field energy; the latter have not yet been reported.

Energy supply systems are classified as follows:

1. direct connection: electric current, electromagnetic waves (especially light waves);
2. indirect or telemetry: electromagnetic waves (especially light waves) (microphotoelectric conversion, ion drag force, PZLT, photothermal conversion, photomechanical conversion); electric field (polymer gel actuator); magnetic field (super magnetic flux actuator, electromagnetic induction); vibration field (forced vibration field, thermal molecular vibration field, sonic or supersonic energy);
3. internal energy sources: chemical battery, solar battery, chemomechanical conversion.

A direct connection system (tethered system) is the most practical energy supply method. We can easily design and produce an energy supply system for tethered micromachines. The most popular method is to use an electric system, because the majority of practical microactuators or sensors are driven by electric power. However, there are some environments where an electric system cannot be used, for example for medical equipment in the human body. For such a situation we can use electromagnetic waves through an optical fibre. A microscopic laser scalpel is a good example. An optical energy supply system would be effective for the elimination of electronic noise from a micromachine system. The development of a microoptoelectronic energy conversion system would be ideal for application as an actuation system in micromachines.

An Ni–H microbattery has been developed as part of the Japanese micromachine project, but optical, ultrasonic, and electromagnetic methods of energy supply remain to be realized.

5
Applications

Sensor technology has been the first field in which micromachines have been applied. In 1990, Howe *et al.* published an article in *IEEE Spectrum* entitled 'Silicon micromechanics: sensors and actuators on a chip' in which they stated 'Remarkable advances in sculpting minute sensors, motors, valves, and pumps from silicon could affect such diverse fields as optical signal processing and magnetic recording technology.' Five years after that report, the situation is unchanged. Silicon is one of the best characterized materials in the world (Table 5.1 (a) and (b)). It is very strong, has a similar modulus of elasticity to that of steel, and lacks mechanical hysteresis. It seems like a perfect material for sensors. As the electronic properties of silicon are sensitive to stress, temperature, and other environmental factors, many silicon technologists have wanted to assemble transducers and electronic circuits on the same chip. Manufacturing and research and development infrastructures for silicon are already well established, as are materials knowledge and availability and processing expertise, built up over more than 30 years of practical experience funded by investment of billions of dollars by the IC industry. As a result of all this work silicon and associated thin-film materials such as polysilicon, silicon nitride, or aluminium, can be micromachined in batches into a vast variety of mechanical shapes and configurations (Bryzek *et al.* 1994).

Outside of silicon technology, technological infrastructure and know-how are still too immature to permit industrial applications. But from the many fabrication techniques still under development, we will consider possible future applications in industry, the domestic market, and medicine (Table 5.2).

5.1 Industrial applications

Micromachine research is just beginning. Research and development in new interdisciplinary areas is required to explore the large number of possible applications. In the course of this research we must find some fields in which micromechanical systems can be reliably applied in order that we can set up some developmental targets. What kinds of applications are expected for micromechanical systems?

Table 5.1(a) Relationships of etchant recipes, their etching speeds and mask materials (adapted from Petersen (1982))

Etchant (diluent)	Typical compositions	Temperature (°C)	Etch rate (μm min^{-1})	Anisotropic (100)/(111) etch rate ratio	Dopant dependence	Masking films (etch rate of mask)
HF HNO$_3$ (water, CH$_3$COOH)	10 ml 30 ml 80 ml	22	40	1:1	$\leq 10^{17}$ cm^{-3} n or p reduces etch rate by about 1/150	SiO$_2$ (300 Å min^{-1})
	25 ml 50 ml 25 ml	22	40	1:1	No dependence	Si$_3$N$_4$
	9 ml 75 ml 30 ml	22	7.0	1:1	—	SiO$_2$ (700 Å min^{-1})
Ethylene diamine Pyrocatechol (water)	750 ml 120 g 100 ml	115	0.75	35:1	$\geq 7 \times 10^{19}$ cm^{-3} boron reduces etch rate by about 1/50	SiO$_2$ (2 Å min^{-1}) Si$_3$N$_4$ (1 Å min^{-1}) Au, Cr, Ag, Cu, Ta
	750 ml 120 g 240 ml	115	1.25	35:1		
KOH (water, isopropyl)	44 g 100 ml	85	1.4	400:1	$\geq 10^{20}$ cm^{-3} boron reduces etch rate by about 20	Si$_3$N$_4$ SiO$_2$ (14 Å min^{-1})
	50 g 100 ml	150	1.0	400:1		
H$_2$N$_4$ (water, isopropyl)	100 ml 100 ml	100	2.0	—	No dependence	SiO^2 Al
NaOH (water)	10 g 100 ml	65	0.25–1.0	—	$\geq 3 \times 10^{20}$ cm^{-3} boron reduces etch rate by about 1/10	Si$_3$N$_4$ SiO$_2$ (7 Å min^{-1})

Table 5.1(b) Characteristics of materials used for micromachines (adapted from Petersen (1982))

	Yield strength $(10^{10}$ dyn cm$^{-2})$	Knoop hardness (kg mm^{-2})	Young's modulus $(10^1$ dyn cm$^{-2})$	Density (g cm^{-3})	Thermal conductivity (W cm^{-1} °C^{-1})	Thermal expansion $(10^{-6}$ °C$^{-1})$
*Diamond	53	7000	10.35	3.5	20	1.0
*SiC	21	2480	7.0	3.2	3.5	3.3
*TiC	20	2470	4.97	4.9	3.3	6.4
*Al$_2$O$_3$	15.4	2100	5.3	4.0	0.5	5.4
*Si$_3$N$_4$	14	3486	3.85	3.1	0.19	0.8
*Iron	12.6	400	1.96	7.8	0.803	12
SiO$_2$ (fibres)	8.4	820	0.73	2.5	0.014	0.55
*Si	7.0	850	1.9	2.3	1.57	2.33
Steel (max. strength)	4.2	1500	2.1	7.9	0.97	12
W	4.0	485	4.1	19.3	1.78	4.5
Stainless steel	2.1	660	2.0	7.9	0.329	17.3
Mo	2.1	275	3.43	10.3	1.38	5.0
Al	0.17	130	0.70	2.7	2.36	25

*Single crystal

Table 5.2   Micromachined products

Industrial
    Smart sensors
    Microparts and fittings
    Microactuators
    Integrated fluidics
    Microgenerators
    Microswitchers
    Microoptical parts
Industrial and domestic products
    Microrobots and microrobot fingers
Medical
    Smart catheters
    Micromachined operation equipment
    Micromachined endosondes

The obvious advantage that a micromachine has over a conventional machine is its small size: small size is ideal for the following applications:

1. In biotechnology and medicine, where microsize machines would be needed to work inside cells and tissues.
2. In harsh environments, such as in a vacuum, under high pressure, and in ionizing radiation, where manned machines cannot operate. Space, the deep sea, and atomic plants are situations where micromachines could be useful.
3. In narrow spaces or narrow channels where conventional machines cannot reach. Safety checking of complicated plant and minimally invasive surgery are good examples.
4. The strongest incentive for use of micromachines may be cost reduction as only a few small amount of expensive resources would be required, especially if the machines can be produced on an industrial scale in batch fashion.
5. Microactuators and microfittings could be used to assemble smaller instruments and softer machines, suitable for use as future small industrial robots and domestic equipment.
6. Small machines use less energy and fewer materials, factors which are important as we aim to conserve the Earth's resources.

The advance of silicon process technology in industry over the past three decades has motivated micromachine development. The first industrial applications have been in sensor technology, as discussed in Chapter 4. The manufacturers of ICs have predicted that the cost of micromechanical parts and elements could be reduced by increasing productivity per wafer and throughput. But future micromachine technology depends not only upon silicon technology but also upon new microtechnologies. In order to develop the microtechnology for widespread industrial and medical uses, many kinds of micromechanical parts need to be systematically developed next stage. Starting from pragmatic silicon microelectromechanical systems, we shall review current developments and future prospects.

5.1.1 Microsensors

Over the past decade microcomputer costs have decreased by many orders of magnitude. As a result, it is now possible to use micromechanical sensors combined with microprocessors in many applications. As the prices of the sensors has decreased by approximately two orders of magnitude we can now use them in many previously impractical fields.

The commonest silicon micromachined sensor is a pressure sensor. Such a sensor has many applications in the motor industry, medical instrumentation, cockpit instrumentation, and many hydraulic and pneumatic consumer products. Originally silicon piezoresistive pressure sensors were developed as manifold absolute-pressure sensors, which were applied to adjust the air–fuel intake ratio, to measure turboboost pressure, and to improve fuel consumption. Silicon capacitive absolute-pressure sensors were also introduced into the automotive market, and the average price of a fully signal-conditioned package decreased by $10. Today, more than 20 million such units are produced every year.

Silicon micromachined pressure sensors are also used as disposable medical pressure transducers. Disposable blood-pressure sensors are used for direct and continuous blood-pressure monitoring for two or three days after an operation. The system was first introduced to prevent viral infection due to the reuse of blood-contact materials, but was found to be very easy to use and to have low maintenance cost. The system cost rapidly decreased to $10, and more than 17 million sets per year are sold in the USA. The concept of an absolute-pressure sensor has been applied to many other pressure measurements in medicine.

Silicon micromachined acceleration sensors are also common in industry. The automotive industry dominates the growth of this technology. Air bags, smart suspensions, anti-skid braking systems, four-wheel drive, and smart engine mounts involve real-time computation of complex algorithms whose inputs are the signals from micromachined acceleration sensors.

Sensing devices with built-in intelligence, which usually means some form of digital signal processing, are called smart sensors. The first smart sensor was developed some ten years ago and opened up the possibility of plug-in transducers. Recently, the incorporation of a manufacturing interface with a local bus into sensor electronics has become one of the newest trends in the technology. This is useful for smart sensor technology, as smart sensors include not only micromachined sensors but also other solid state sensors. The IEEE task force TC-9, in cooperation with the National Institute of Standards and Technology (NIST), recently began the development of a standard for low-cost smart sensor communication. The first Smart Sensor Communication Workshop was held on March 31 and April 1, 1994, at the NIST office in Gaithersburg, Maryland, USA, and discussed digital signal formats and network interface capability.

The most delicate and complicated microelectromechanical system is the probe of an atomic force microscope (AFM). The top of an AFM stylus tip is fabricated by dry etching, at close to atomic scale. The tips are installed on cantilever beams processed using surface micromachining combined with bulk micromachining of the silicon substrate and integrated with sense and drive electronics (Kong *et al.* 1990). The scanning mechanism is also microfabricated by silicon micromachining technology. The scanning drive mechanism is modified from a crab-leg flexure resonator, or comb resonator, whose resonant frequency changes with acceleration, pressure, or other physical phenomena. For example, a laterally driven microbe tip for use in a scanning thermal profilometer was developed by University of Michigan using a bulk dissolved-wafer process in combination with reactive ion etching (Gianchandani and Najafi 1991).

5.1.2 Micromechanical and microstructural devices

The market for micromechanical devices is small at present. Combined techniques of bulk and surface micromachining have been used for most of the silicon micromechanical devices. In bulk micromachining, the wafer is processed by a bulk dissolved-wafer process including the use of impurity-based etch stops (Chaw and Wise 1988). Some advanced neuroelectronic interfaces have been processed using this technique and practical applications include a multichannel multiplexed intracortical microprobe with an integrated silicon ribbon cable (Najafi *et al.* 1985; Najafi 1994), a three-dimensional microsystem consisting of multiple neural probes assembled in a micromachined silicon platform (Hoogerwerf and Wise 1991), and a micromachined silicon nerve regeneration electrode (Akin *et al.* 1994).

LIGA and SLIGA (LIGA combined with the use of sacrificial layers) have been adapted for use with this technology. Products include thermally actuated microrelays, micromotors, magnetic actuators, microoptics, magnetic actuators for relays, voltage regulators, valves, and microconnectors, composed of many micromechanical components like joints, springs, bearings, gears, conduits, and nozzles. LIGA services are already being offered by companies such as IMM (the Mainz Institute for Microtechnology) and Microparts, both in Germany.

5.1.3 Future trends in industrial applications of micromachines

The energy conversion system is a key component which moves robots and manipulation arms. Actuators or engines such as micromotors, microturbines, and micropiston engines need to be developed as a first stage. Microelectrostatic rotating motors the diameter of a human hair, wobble harmonic motors with off-centre rotors, linear microactuators similar to muscle sarcomere, and micro-Stirling engines (a kind of heat engine) are examples of actuators that have been developed and tested for micromachine energy conversion systems.

As mentioned in Chapter 4, one problem is how to supply energy to microactuators. Microbatteries need to be developed to supply electric power to microactuators, but how do we make such microbatteries? Paper electric cells and solar cells have already been miniaturized, but they do not have sufficient power to supply microsystems. Temperature differences are used to supply energy to Stirling engines. Chemical energy is an ideal energy source for a microactuator, but some chemomechanical energy conversion system needs to be invented. Other field energies, such as electric field, supersonic and electromagnetic field energy might be better energy sources for microactuators, because they can be situated externally, outside the microsystem. Microenergy sources are the first priority for micromachine development.

When we make a microactuator system from many microparts, control signals need to be supplied through electric lines and microconduits. In the fabrication of recent integrated circuits, we still face a problem in making submicrometer wires on a chip; the energy and signal transmission systems could be similar bottle-necks to development in micromachine research.

Since a microrobot is an autonomic mobile machine, it must monitor its environmental conditions, recognize the situation, and decide on its next action. This means that micro size sensors (not microsensors) and micro size logic circuits are indispensable for microrobot development. Will logic circuits reach such a small size in the near future?

When a machine is working in the human body or other enclosed spaces without communication lines, we must detect its location and information on its status. A purpose-made communication system would be needed in a microenvironment. Such a system does not exist in biology (except chemical communication), because living organisms do not require external monitoring of their bodies. We therefore do not have any biological models which we can use for the development of micromachine communications.

Considering micromechanical systems systematically, micromachine research includes many items which must be developed in the near future. Some spin-off products from micromachine technology have already contributed to the development of microactuators and microsensors, resulting in the production of new types of measurement instruments. Micromachine parts will have huge application in industry and medicine.

5.2 Micromachines in the home

Domestic electrical products have a large market in industry. Micromachined parts are used in video players, 8 mm home video cameras, personal computers, and many electronic integrated circuits. In this respect, today's consumer products can be regarded as micromachines. However, it is difficult to imagine what real technological breakthroughs could mean for the application of micromechnical devices in consumer products. Future consumer products produced by micromachine technology are likely to belong to the following four categories:

1. micromachines for entertainment;
2. domestic microelectronics products;
3. domestic maintenance microrobots;
4. environmental control robots.

Today, insect robots have been developed by artificial intelligence researchers and some of them are sold as toys or for educational purposes. In 1989, Flynn, at MIT Artificial Intelligence Lab, reported a 1 in^3 autonomous robot; it had two wheels and drove two microelectromagnetic motors, a battery, and a neural network controller. It stood on a table and hid in a shelter. When the controller whistled, the mechanical insect would come out of the shelter in response. Robots named 'Mr. Monsieur', produced by Seiko-Epson Co., have been sold as an educational tool for teaching the group control mechanism of autonomous distributed robots. The driving mechanism of the robot is constructed from vibration actuators and mechanical parts of wristwatches and the robot is navigated by a light signal. The external size is only 1 cm^3.

Similar insect robots have been made using many lithographic techniques, and robots driven by shape memory alloys have become the most common type.

5.3   Medical applications

Many fields of medicine involve work on a microscopic scale, because the physical environment and dimensions used in the study of cells, organs, and bodily fluids are on a very small scale. However, medicine has lacked the perspective of mechanical systems that is common in engineering. Research has concentrated on analysing the behaviour of single protein molecules and organelles, or the macrolevel functions of tissues. It must be remembered that living organisms have up to now not been considered as machines.

In modern industry, machines can only be mass produced when there is a large supply of a wide variety of interchangeable components. Thus, there are machines available at present which contain components of 1 mm or less, but they are not so much modern machines as remnants from craft industries of the past. Examples would include many micromanipulation tools in medicine. These small mechanical components may sometimes reach a precision of 1 μm or so, but they all require large machine tools to produce. This is not the kind of technology that micromachines are concerned with. Micromachines require standardized groups of components which include micromechanisms. In this sense, they are more closely modeled on the system components produced using semiconductor technology.

Even if small components are available, the technology to combine them must also be there. This will only become possible through the development of batch pre-assembly (like the silicon process) and molecular autoassembly (like protein technology), and the microprocessing machinery and software to assemble them.

5.3.1 Micromachines for medicine

We cannot be sure of the direction which future technological advances will take in medicine. But when we classify the possibilities under the keywords material, information, and energy, every technological trend indicates that future technology aims at working on supermolecules, and cells instead of organs. Such techniques are usually considered as 'biotechnology'. We must first define the application of micromachine technology from the biotechnological standpoint, and secondly, mention micromachine applications in clinical medicine generally.

Trends in biotechnology

We classify biotechnology in three subgroups the second of which can be further subdivided by combining materials and processes with biological and artificial methodology (Table 5.3). Each subgroup has its own specified demand for micromachine technology.

1. **Type I: biotechnology (in a narrow sense).** Biotechnology usually means the manufacture of useful products from natural resources using cells as 'black boxes'. Brewing using fermentation is typical of traditional biotechnology. Cellular engineering and genetic engineering have introduced new techniques by which we can artificially change the genetic information of a cell. Key mechanical techniques of biotechnology are micromanipulation for cell handling and injection

Table 5.3   Four types of biotechnology

	Materials	Production techniques	Products
Biotechnology	Biological	Cellular engineering Genetic engineering	Useful bioproducts
Biomimetics	Artificial	Simulation of function	Artificial organs, robots
Biological machines	Cells, organelles, supermolecules	Self-organizing	Hybrid artificial organs, biocomputers
Artificial biomechanism	Artificial	Autoassembly	Artificial structural materials

of chromosomes or DNA fragments into a cell. The size of such objects is usually less than 100 µm. Micromachines such as laser tweezers (Block 1992) or integrated microfluidics (Kabata, 1993) are good examples of micromanipulators used in this field.

2. **Type II: New biotechnology: biological machines and artificial structural materials.** In order to explain 'new biotechnology', we must consider one further type of biotechnology. This technology depends upon ordinary engineering and produces machines such as artificial organs and robots from non-biological materials to imitate functions of the living body. We refer to the technology as type III.

New biotechnology (type II) can be defined as the method of production of biological machines (type IIa), and the technology for the fabrication of materials to make artificial structures (type II b). These two technologies are mainly relevant to nanotechnology, but we shall describe them briefly here.

In biological machine technology (type IIa), we use cells, organelles, or super molecules as materials and fabricate hybrid artificial organs, biocomputers, or biological robots. The key scientific knowledge and techniques required will be the self-assembling mechanisms of mesoscopic organic materials. Even a decade ago we could not imagine such techniques, but today we can find a few examples. Controlled cell proliferation on inorganic materials, especially on silicon wafers, has been tried and reported by neuroscientists and computer engineers. They intend to form communication networks between a cultured neural network and an electronic network (Jimbo and Kawana 1992). Hybrid artificial organs have been developed using the same methods by which cells have been cultured on layers of artificial materials. Artificial liver and pancreas have been studied clinically. Even though striated muscle was used as the actuator of an artificial heart, some artificial hearts have been tested clinically. It is certain that tissue engineering and organ fabrication will be possible in the future (Amato 1992; Langer and Vacanti 1993).

Artificial biomechanisms or artificially structured materials (type IIb) represent another type of hybrid biotechnology. Source materials used for the technology are principally artificial and not biological materials, but the products have the same structure as some living organisms. Production will be by autoassembly system, but remains to be developed. Examples of such products are Langmuir–Blodgett bilayer membranes, nanotubes which are modified fullerene tubes (Iijima and Ichihashi 1993) or peptide tubes, and artificial sarcomeres (Fujimasa 1992). 'Molecular tectonics' has been described in crystallography (Mann 1993), and in the near future, we could use supermolecules produced by this process to build micromachines.

3. **Type III: biomimetics or artificial biomechanisms.** Biomimetics and artificial biomechanisms are closely related micromachine technologies. The technical principle depends on imitating the physiological functions of a living bodily system or an organ. The source materials are usually inorganic and

products are typical machines whose structure is *not* analogous to organs or biological systems. The most important point about these products is that the machine has almost the same physiological functions as the simulated organs. Many artificial organs, robots, and computers typically simulate the function of bodily organs, hands and arms, and brains. Therefore, the design principle is based upon physiology; on the contrary, that of an artificially structural material is based on anatomy. As an example, when aeroplanes were developed they did not imitate bird anatomy, but did operate on the principle of bird's flight. As this type of biotechnology is based upon macroscopic function, ordinary engineering is applicable and the micromachine technology can be adapted to future demands.

Trends in clinical medicine

In biotechnology, micromachine technology is still at a basic level, but in general clinical medicine, many practical applications of micromachines appeared during the 1980s (Table 5.4).

Micromachine technology is well suited to the production of machines with medical applications: machines which act on the body or its cells. This technology

Table 5.4 Future trends in medical technology: measurement, treatment, systems, and information

Technological domain	Trends of technology: KEYWORDS
Measurement	
Observation	from VISIBLE to INVISIBLE
Backgrounds	from PHYSIOLOGY to ANATOMY
Objects	from BLOOD to CELLS
Substance	from METABOLITES to GENES
Treatment	
Methodology	from MACRO to MICRO
Operation	from DIRECT to INDIRECT and REMOTE
Sensing	from REAL to VIRTUAL
Fields	from SURGERY to INTERNAL MEDICINE
Techniques	from INVASIVE to MINIMALLY INVASIVE
System	
Places	from HOSPITAL CARE to HOME CARE
Action	from THEIR to SELF
Status	from CONCENTRATED to DISTRIBUTED
Principles	from ORDER to ANARCHY
Information	
Theory	from UNIFORMITY to CHAOS
Information	from CENTRALIZED to AUTONOMOUS and DISTRIBUTED
Evolution	from DNA to MIEME
Interface	from EXTERNAL to INTERNAL

is concerned with nothing less than the creation of components able to work on the structure and functions of systems, cells, and other units far smaller than bodily organs. Medicine already makes use of machines that have dimensions in the submillimeter range, but these do not work inside cells, tissue systems, or fine blood vessels. Completely new production methods and concepts are required to operate with such fine targets.

The two most common fields of application at present are surgery and pharmaceuticals. Micromachines will mean a drastic shake-up in medicine. Already the world of 'old medicine' is undergoing a shift, as micromachines, virtual reality techniques, tele-operation, and intervention change traditional surgical techniques towards those used in internal medicine. Physicians, meanwhile, are developing new treatment methods—minimally invasive surgery and internal robots—in an effort to promote localized drug delivery systems. But these goals will not achieved without the development of micromachine technology. This market accounts for over one-third of total medical expenditure. It is vital that micromachines be utilized in this area first.

It is likely that microsystem technologies will be applied in the six categories in future biomedical engineering fields (see Table 5.5).

Table 5.5 Micromachine applications in future biomedical engineering

Medical applications	Future technological seeds
Minimally invasive surgery	
Endoscopic surgery	Virtual reality, remote operation system, microrobotics
Laparoscopic surgery	Stereo vision, remote microsensing system, virtual reality
Laser angioplasty	Fine fibre optics, microactuator, nanometric sensing system, virtual reality
Computed surgery	Visible human project database, stereotaxic operation, computer-guided operation instruments, simulated patient images (surgical planning, micro custom parts production)
Microscopic surgery	Televised microoperation, virtual reality, microrobotics (catheter microscope)
Biotechnological applications in medicine	
Implantable artificial organs	Microactuators, micro energy sources, microparts, microsensors, autonomous distributed control method
Drug delivery system	Nanoactuators, artificial membrane technologies, remote control technologies for microparticles
Micromanipulation	Submicrometer mechanical probes, cell handling technology, integrated fluid circuits

The demand for micromachined products clearly exists in minimally invasive surgery. But for some of these applications, conventional production techniques are still somewhat of a craft. We can assemble some of these machines as microelectromechanical systems, but most of the medical micromachines include mechanical, chemical, and optical systems as component parts. We will discuss the present status and the technical trends of micromachine applications in medicine separately.

5.3.2 Minimally invasive surgery

One more important medical application of micromachines is for surgery in areas of the body, where organs and tissues are invisible to the naked eye and impossible to touch hand. Usually such surgery is called 'video assisted surgery (VAS)', 'keyhole surgery', or '*Nintendo* surgery', and combines the use of micromachine, multimedia, and computer technology. The techniques of VAS depend upon remote surgical operation through narrow channels, and we can operate less invasively as a consequence of it. Recently such procedures have generally been called minimally invasive surgery.

The word 'keyhole' implies finding some pathway through non-vital tissues or organs for approaching the site requiring surgery. Initially, some silent areas of brain were used for stereotaxic surgery, but today some transcutaneous direct endoscopic surgery has been called keyhole surgery. Laparoscopic cholecystectomy, sterate gangrionectomy, myomectomy of the uterus, lobulectomy of the brain, etc., are now usually performed in 'keyhole' fashion. By the introduction of stereoscopic endoscopes and microrobotic hands, almost all minor abdominal and thoracic operations will be performed through the keyhole.

Minimally invasive surgery is categorized into four methods. One is surgery of the digestive tract and other hollow or tubular organs with the assistance of an endofibrescope. This is usually called endoscopic intraluminal surgery. The second category is 'keyhole surgery' *per se*. Surgeons open small holes in the skin and operate using a remote operative tool through the 'keyhole'. The third category is transvascular surgery, in which arteries and veins are used to guide cannulae. The final category is stereotaxic surgery, which has been used in brain surgery and has recently spread to interventional surgery using medical imaging. Computer assisted surgery (CAS) or computed surgery (CS) belongs to this category (Taylor *et al.* 1996).

Remote sensing and manipulation through narrow channels are necessary for these surgical procedures, and we call such surgery 'microchannel surgery' (MCS).

Endoscopic intraluminal surgery
Endoscopic intraluminal surgery has become familiar to both surgeons and physicians, and the techniques are relatively well established. Gastrointestinal, rectal, tracho-bronchial, vaginal, urethral, and intracranial endoscopes have been

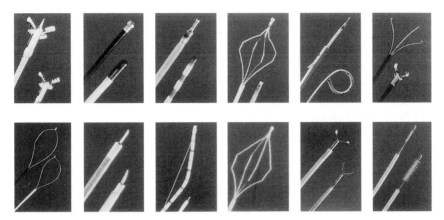

Fig. 5.1 Various remote manipulating instruments for fibre endoscopic operations. (From catalogue of endoscopes by Olympus Co.)

used. Generally, these procedures involve parts of the body which have an external orifice. Thus permitting man-made equipment to be introduced without the need to consider biocompatibility. Minimally invasive surgery began as a natural result of endoscopic intraluminal surgery.

An endoscope is equipped with an an imaging fibre, light sources, and a working channel, and its external diameter is less than 1 cm. Many miniaturized remote operation tools (Fig. 5.1) have been developed for insertion through the working channel of an endoscope. The endoscope tip is usually able to bend, and is controlled remotely by sheathed cables. Today some endoscopes are designed to contain even thinner endoscopes which can be introduced by the surgeon into a narrower side duct such as the choleduct or gallbladder and be manipulated separately.

There are many kinds of applications for endoscopes in clinical medicine but there are also many more opportunities for developing more effective endoscopic surgery. Many micromanipulators used in a remotely manipulated catheter, such as scissors, tweezers, knives, clamps, grippers, etc., are disposable. These tools have been fabricated by craft methods and are therefore expensive. Some powered micromanipulators, such as grindstones and drills, are actuated by thin coiled wires powered from outside the endoscope. This system is fine but difficult to manufacture. When we are able to apply micromachining techniques to such micromanipulator fabrication, we will get microdisposable tools of comparable cost to today's disposable knives and syringe needles. Surface micromachining can construct microgrippers and wobble motors actuated by electrostatic forces, and such microdevices will be of use not only in endoscopic surgery but also in microsurgical techniques.

One more problem encountered when using an endoscope is how to guide it to the right place. In the digestive tract, the surgeon must guide the endoscope through a very soft and winding channel, and in the case of the large intestine, against the prevailing direction of intestinal peristalsis. From the very early stages of micromachine projects, the development of an automatic endoscope guiding system was planned. Such an endoscope, commonly called an active endoscope, has segmental structures along the catheter. Each segment connects to a fine shape memory alloy (SMA) actuator, which bends the catheter by heat deformation. For application in the large intestine, the endoscope was designed with the same outer diameter as conventional endoscopes (13 mm) and was able to pass through the sigmoidal intestine with ease (Ikuta 1988*a*). Application of SMA actuators to endoscopic catheters has many benefits, because the SMA element can detect its own deformation status from the temperature of the element measured as the resistance.

Laparoscopic surgery and fiberscopic surgery

One of the modern trends in surgery is the development of less invasive operative techniques. Instead of conventional surgery, remote operation using an endoscope or other intervention techniques is becoming more wide spread. The key enabling technology is the optical endoscope system which is usually used for observing abdominal organs through a hard, straight tube (penetrating the abdominal wall), at both ends of which are placed optical lenses. The hard endoscope is called a 'laparoscope', 'thoracoscope', and so on depending on the part of the body being observed. This system has many micromachine applications.

Laparoscopic surgery is an old surgical procedure and was applied in urogenital and gynaecological surgery in the early 1900s. Modern laparoscopic surgery began after the introduction of a televised laparoscope. Since the late 1980s televised operations using a hard laparoscope and a laser scalpel have been performed. Laparoscopic surgery to the myoma of the uterus was the first major application, but the technique has been developed for cholecystectomy (Reddick and Olsen 1989; Dubois 1990; Perissat 1990). Advantages of laparoscopic surgery are as follows:

1. It is a minimally invasive method.
2. There are fewer post-operative complications, especially low adhesion of the peritoneal membrane.
3. There is a shorter stay in hospital.
4. There is almost no scar on the abdominal wall.

It might be considered that the most important innovation in laparoscopic cholecystectomy is the systematic introduction of micro disposable parts into surgery: the micromechanical parts such as scissors, tweezers, knives, and grippers, were developed as disposable and interchangeable parts of remote manipulators; staplers for stopping bleeding and trocars, which guide the manipulators into the abdominal cavity, are also disposable. The laparoscope and laser scalpel are nondisposable parts of the system. These parts are systematically arranged and partly mass-produced as micro disposable units (Fig. 5.2).

This system suggests to us the surgery of the future. Most operations will be performed inside the body cavity through narrow channels in the body wall. New types of remote microoperation techniques will inevitably appear. What is the surgical breakthrough? Let us examine the procedure in laparoscopic surgery.

(a)

(b)

(c)

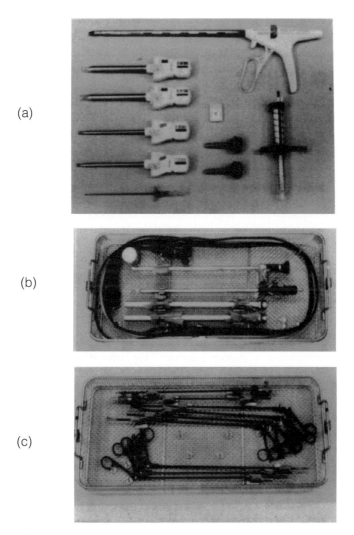

Fig. 5.2 Instruments for laparoscopic surgery: (a) a surgical stapler and trocars; (b) a fibre scope and laparoscopes; (c) remote operating instruments (scissors, tweezers, clamps, etc.). (Adapted from Idezuki (1989).)

Recent laparoscopic cholecystectomy uses four or five trocars inserted through the abdominal wall at one time for guiding many straight instruments. A trocar is a hard, straight hollow tube with an external diameter of almost 6 mm to 11 mm. The endoscope is a hard straight laparoscope. Images of the abdominal cavity are sent by a lens system to a television camera via charge coupled devices (CCD). Almost all inserted instruments used in the operation are hard and of a rod-like shape; this is because the theatre staff must hold organs and tissues by these instruments indirectly, and it is impossible to hold and keep them still from certain directions. Therefore, trocars are inserted in several parts of the abdominal wall to guide the various remote manipulating tools. During the operation the abdominal cavity is inflated with low-pressure carbon dioxide gas to provide a clear field of view. Some surgeons pull up the abdominal wall with wires and make a tent-shaped space in the abdominal cavity (Hashimoto 1995). A chief surgeon watches televised images of the organs and tissues produced by a laparoscope. The operational procedure is the same as with open abdominal surgery, except that the surgeon's hands and fingers cannot enter the body cavity. A laser scalpel is used for cutting tissues. The gallbladder is pulled out from a small incision, tied and cut off at its connection to the bile duct. The abdominal wounds are very small and almost no scar remains on the skin afterwards. For comparing the prognosis of laparoscopic cholecystectomy and open abdominal surgery, a patient's quality of life after the operation is a good indicator. Usually patients who have had laparoscopic surgery can leave hospital two days after the operation but the other patients cannot leave for a week or more. The former can eat foods on the first day after the operation without feeling pain, whereas the latter cannot eat for two days or more and need strong sedatives.

Why did such simple operational procedures not prevail before 1980? Patients always hope for a less invasive procedure but surgeons usually want to operate under as broad a field of manipulation as possible. Moreover, surgeons usually prefer to use a 'hands-on' approach to prepare and manipulate instruments, tissue, and organs. To paraphrase a famous surgeon: 'The operation is half-way to success when we can obtain a broad enough field of view. The patient suffers from some pain afterwards, but it doesn't hurt me!' Lack of understanding of the discrepancy between patient's and surgeon's views has prevented the improvement of less invasive techniques. If we want to replace conventional surgical processes with less invasive methods, we should try to maintain the conventional theatre environment as much as possible. Even after establishing a minimal invasive surgery system for a certain operational technique, surgeons may still ignore patient's opinions. The rapid popularization of laparoscopic cholecystectomy is due to the fact that the laparoscopic surgery system can afford almost the same surgical environment as conventional cholecystectomy.

The early picture of micromachine developments indicates that some smaller and softer operating instruments might be devised. As an example, if inserted instruments could be developed with soft materials and be able to bend, and if these instruments could work as a robot hand does, the system would be changed completely. The operation would require fewer trocars. However, many experienced laparoscopic surgeons who have used soft fibrescopes have said that it is more difficult to orentate soft fiberscopes than hard laparoscopes. The largest problem is that a surgeon cannot realize a tip position and direction of an operation tool in the abdominal cavity if he uses a fiberscope to watch the tool, because he cannot know the direction of the fiber there. For preventing such problems, we must develop micronavigation systems and positioning sensors such as a magnetic field positioner, micro-gyroscope, or microaccelerometer on the tip of the endoscope.

The application of stereovision will improve the system. The anastomoses of the intestine have already been viewed in clinical cases under stereovision using a new stereo-laparoscope with a double optical system in a trocar (Hashimoto 1992). The technique will inevitable lead to the development of small remotely controlled instruments for small abdominal operations.

For such a remote operation, many sensing transducers will be required to dectect tissue characteristics and the physical and chemical environment, a job normally performed by the surgeon's many off-line sensors. A miniaturized electronic scanning ultrasonic detector has been developed and used to differentiate string-like tissues into choleduct, arteries, veins, nerves, or connective tissue (Hashimoto 1992).

Laser angioplasty

All tissues and organs are supplied with nutrients and oxygen by their arterioles and dispose of their products or waste materials and carbon dioxide through their venules. Therefore, if we can guide a fine endoscope through the arterioles or venules, we can approach and treat almost all organs and tissues.

Since 1974, many vascular surgeons and physicians have developed many kinds of instruments for angioplasty (Gruntzig and Hopff 1974). Transluminal coronary angioplasty (TCA) is the one of the most successful technologies applied to coronary occlusive diseases today. Percutaneous transluminal coronary angioplasty (PTCA) is used to treat 400 000 people per year in the USA at a cost of over $60 billion per year. Laser angioplasty and fiberscopic laser coronary angioplasty have been developed and tested on clinical cases, and direct imaging excimer angioplasty has been a great success for repairing obstructions in coronary arteries. In brain surgery, endovascular treatment of intracranial aneurysms has been tried, and it is planned to introduce micromechanical technologies into the field.

In the coronary angioplasty, the lumen of a narrowed coronary artery can be enlarged using a balloon catheter. But a balloon catheter cannot be used when the artery is completely clotted or the patency of the vessel cavity is doubtful. Laser angioplasty without an imaging system has been developed for such cases (Abela *et al.* 1985; Myler *et al.* 1987; Sanborn *et al.* 1988). These blind laser angioplasty methods have been criticized because of the risk of perforation of the vascular

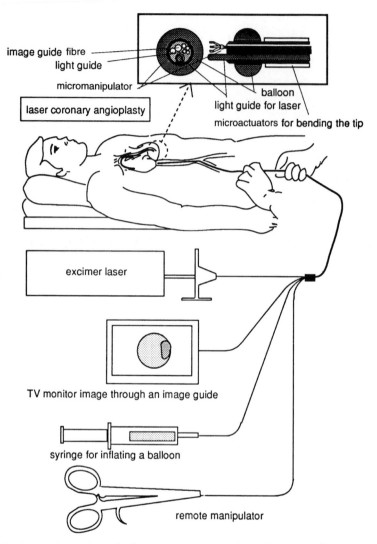

image guide fibre
light guide
micromanipulator
laser coronary angioplasty
balloon
light guide for laser
microactuators for bending the tip

excimer laser

TV monitor image through an image guide

syringe for inflating a balloon

remote manipulator

Fig. 5.3   A complex catheter for laser coronary angioplasty. A 0.4 mm diameter quartz optical fibre for the ultraviolet laser, a 0.4 mm diameter image fibre with a bundle of 4000 fibre elements for observation, light guide fibres, a channel for the passage of liquids, and balloon inflation and deflation channels are include in the catheter which has an external diameter of 1.5 mm. The tip of the catheter can be bent with two degrees of freedom and controlled by two remote hand manipulators.

wall. To overcome this, laser angioplasty with an ultrathin image fiber has been developed. As the external diameter is limited to 1.5 mm at most, many technological demands exist for the development of micromachine technology.

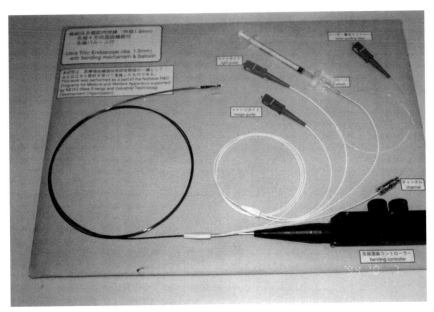

Fig. 5.4   A catheter for fibrescopic laser angioplasty, which includes a laser quartz fibre (0.4 mm diameter), an image fibre (0.4 mm diameter with 4000 fibres), a balloon channel, and a working channel with a remote tip bending tool.

Improvement of image quality would be the first target for microsystem technology (Figs 5.3 and 5.4). As angioscopic images are obtained through narrow communication channels, the number of picture elements has been 5000 pixels at most. For example, the outer diameter of a modern finest-image fiber is only 200 μm, but the number of image elements is only 1600 pixels if we use quartz fibres. When clinicians diagnose a pathophysical abnormality using endoscopic images, this number of pixels does not seem adequate. Therefore, we must develop some image enhancement systems. One answer would be optical scanning methodology and another would be stereovision systems: these techniques would provide some supplementary images to the original pixels and create a helpful optical illusion in much the same way as dynamic vision in cinema and television images (Fujimasa 1992; Fig. 5.5).

The ability to remotely bend a catheter tip is a great target for microtechnology. Angioplasty catheters are usually bent using long thin cables. If we were able to use some microactuators on the tip of the catheter these could decrease the cannula diameter and cause remote bending. Recently Toshiba have developed a micromotor, of external diameter less than 1 mm, and a tube having an external diameter of only a few micrometers, able to be bent pneumatically. These actuators may be applicable for microcatheter techniques.

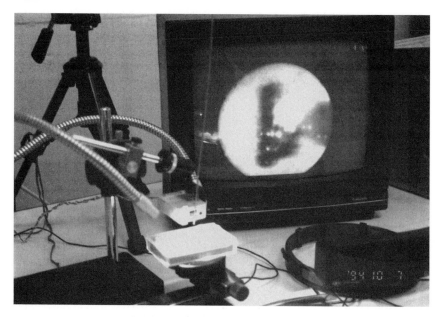

Fig. 5.5 A monofiber stereoscopic image system for remote microoperation. The fibre of the experimental system is forcibly vibrated and two positional images are obtained in opposite positions to that of the vibration wave. Two images are sent to goggles with a liquid crystal image shutter.

Sensing using catheter-immersed transducers is another good example of an area where microtechnology could be applied. In cardiac catheter testing, pressure, blood flow rate, blood gas information, and blood chemical analysis data are collected by transducer systems outside the body. To collect such data in on-line fashion, we must develop transducers able to be mounted on a catheter tip. The size of ordinary electronic transducers is usually 1 mm or more. Furthermore, the electric current passing through the vessels can sometimes cause a microshock to the heart. From that point of view, an optical fiber sensing system would be a better solution (Schultz 1991). Today, a glucose sensor using thin optical fibers which are coated by immunofluorescent materials has been developed, and a pH sensor of similar type has been reported (Barnard and Walt 1991) (Fig. 5.6). The technology could be developed for futuristic cellular sensing devices consisting of just a single fiber (Tan *et al.* 1992) (Fig. 5.7).

Flow guided microangiocatheter   The modern angioscopic catheter was developed by Seldinger in 1957. The catheter can be inserted in the vascular system without the need for an incision. The method is frequently applied in interventional surgery. The finest angiocatheter is less than 1 mm in external diameter and can contain much micro disposable equipment.

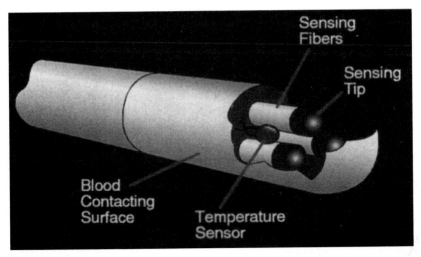

Fig. 5.6 An intra-arterial blood gas (IABG) sensor composed of fiber sensors which are optically irradiated and measured by luminescence. The sensor is immersed in a catheter with a temperature sensor. We can measure pH, pO_2, and pCO_2 in arterial blood continuously for 72 h using the IABG catheter sensor. (From the catalogue of the IABG sensor by Nihon Kohden Co. Ltd.)

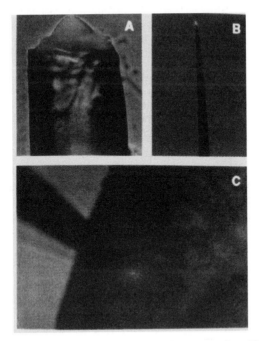

Fig. 5.7 A submicrometer optical fiber pH sensor (B) was produced by drawing out a 105 μm diameter fiber (A). The optical fiber sensor was inserted into the extraembryonic cavity of a rat embryo. Blue light emanates from the sensor (C). Recently, many optical fiber sensors have been developed using evanescent light detection. (From Tan *et al.* (1992).)

For vascular surgery, vascular surgeons and physicians have developed many microtools to use in angioplasty. As the main purpose is to recanalize the blood flow, the most important instrument is a scalpel to incise the thrombus in the peripheral vascular system. In ordinary surgery, steel and electric scalpels are used, but these scalpels cannot be used in the vascular system. Therefore, many kinds of laser scalpel are replacing ordinary scalpels in angioplasty. Direct laser light abrasion and cutting by hot laser tips attached in front of the laser fiber have been used. When the target artery is partially thrombosed the physician can enlarge the narrowed artery using a balloon. The method is usually applied to coronary artery thrombosis, so it is called percutaneous transluminal coronary angioplasty (PTCA). A balloon catheter is routinely used in outpatient clinics as the treatment of first choice for heart infarction. Injection of thrombus-dissolving drugs into the thrombosed artery is also a treatment of choice. The catheter therefore has a guide tube for flushing; the flushing channel is used to guide many microtools and microparts.

In some surgical applications thrombus *formation* is required. In order to prevent brain haemorrhage, brain surgeons make thrombi form in small aneurysms of the brain artery. Thrombus-forming therapy is also used to destroy some malignant tumours. For these applications the catheter tip needs to be sent through complex arterial channels and must transport thrombus-forming materials into the aneurysm or terminal arteries. Guiding micromechanisms have been developed in some micromachine projects. A commercially available guiding catheter has a miniature balloon kite on the tip and drifts down the bloodstream from the femoral vein to the pulmonary arterioles. The kite can open or shut by means of the inflow or outflow of fluid.

Stereotaxic surgery and computer-assisted surgery

Stereotaxic surgery was first introduced to brain surgery, but today we can operate on some other organs using an interventional method assisted by many imaging modalities such as MRI, XCT, and ultrasonic imaging. As the system uses computer simulation methodology it is called computer assisted surgery (CAS). This methodology also requires many miniature probes to locate and operate at the foci.

In stereotaxic brain surgery, the patient's head is fixed with a skull clamp such as a Mayfield clamp. As individual differences in skull and brain structure are relatively small, the positions of important functional foci in the brain are decided deterministically in three-dimensional space. The surgeons introduces an operational probe, which is fixed to the clamp like an industrial robot arm, through a silent area of brain into the diseased part and resects the tumour or removes the blood clot. Recently, some micronavigation transducers have been developed. As an example, a small magnetic field sensor is attached to an operational probe and a magnetic field source is fixed with a skull clamp. The position and the angle of orientation of the sensor were measured with a three-dimensional digitizer using magnetic field modulation. Then the position of the probe tip was calculated by a computer and displayed on the CT or MRI image planes with a cursor. The surgeon removes the tumours with reference to the resection areas mapped on the patient's tomography images (Kato *et al.* 1991).

Clinical trials using the same concepts have been done in interventional surgery of hepatomas. Computer simulation models of the patient's upper abdomen were calculated from MRI images and the direction of piercing by a laser scalpel was decided using a computer.

This technique could be modified in the future to give complete computer controlled surgery, in a way similar to numerically controlled machining. The operation will be performed by robot probes which will be controlled by computer, and surgeons will watch the probe moving on a monitor screen. For such computer-controlled surgery we will need many kinds of small robot 'hands' which will incorporate many micromachine parts.

Micromachine techniques in microchannel surgery

Remote operation through transcutaneous thin conduits, usually called trocars or catheters, is one of the best solutions for less invasive surgery. In order to design a total package of less invasive surgeries, remote operating systems using stereoscopic endoscopes have been developed and tested on animals.

There follows an analysis of this system from the standpoint of micromachine technology, and we include some proposals for improvement in anticipation that we will be able to use some micromechanical parts.

Searches for technical improvements in all the surgical processes discussed so far have a common goal. Communication and operating tools developed using micromachine technology will become indispensable to the processes. Microscopic teleoperation technology will also be important.

System design concepts   Transcutaneously teleoperated surgery consists of the following three main components:

1. **Visual system:** A stereoscopic fiberscope with 6000 imaging filaments enclosed within an external diameters of 0.6 mm was used for stereovision. The image was used for sewing tissue and inserting catheters into blood vessels. An endoscope, which was modified from a laparoscope, was used for rough manipulation during surgery. Images obtained from the laparoscope were displayed on a conventional TV monitor. A head-mounted display for stereoscopic visual images was assembled from two 0.7 in liquid crystal televisions each with 100 000 pixels.

2. **Thin transcutaneous conduit:** The trocars, temporary conduits to guide operating instruments into the body and seal tightly from outside of the body to inside organs or tissues, were modified from those used in laparoscopic operating systems.

3. **Manipulators:** Basically tweezers, forceps, scissors, staplers, a laser scalpel, and an electric scalpel are prepared as for laparoscopic surgery. These manipulators are not sufficient to perform complicated operations. In order to anastomose blood vessels and intestines or to put some stitches in an organ, forceps were used as a needle holder, but remotely manipulated sewing machines will be developed. Retractors, with which obstructing tissues and organs are held aside, were not developed, but some kind of retractor will be developed in the future.

The development of general microscopic teleoperation systems must be guided by one following questions:

1. Is stereoscopic imaging useful for a delicate operation?
2. What types of instruments should we develop for use in a teleoperation?
3. Do we require some new techniques to maintain the teleoperative environment?
4. What types of tools are to be controlled by a stereotaxic procedure?
5. What types of elements do we need for micromechanical parts and machines?

The answers to these questions can be obtained by experiment. Stereoscopic imaging was found to be essential for sewing, but the most important imaging function was working out the orientation of manipulation tools. The field of view of microscopic images is small, and if the manipulator head is lost from view it can be very difficult find the manipulator again. High magnification and low magnification are required at the same time, and stereotaxis is essential for the system.

As the operation is performed through narrow conduits, the shape of every manipulator tip needs to be transformed after entering the body cavity. For the transformation, strings, wires, and springs are usually used, but if we can produce microactuators with enough power many kinds of manipulation tools can be designed. Retractors must be developed, but some conceptual changes will be required.

Anyway, the system is at a very early stage, but many technological demands are becoming apparent. Once the system has been developed—and at present it is still too complicated—it will have a strong influence on surgery.

Microscopic surgery (microsurgery)

Microsurgery is the most typical application for micromechanical systems. Microsurgery is a short form for 'microscopic surgery' and means a delicate operation under a stereomicroscope of at least box magnification. An artery or a nerve with a diameter of 600 μm can be anastomosed and sutured by a well trained operator using 20 μm diameter needles and strings. Ophthalmology, otology, peripheral vascular surgery, neurosurgery, and plastic surgery are common fields for the application of microsurgery. The size of the smallest operable arteries and nerves depends upon the working gap between the object

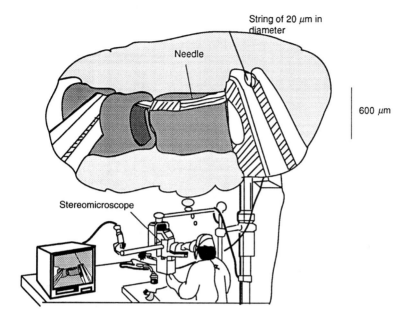

String of 20 μm in diameter

Needle

600 μm

Stereomicroscope

Fig. 5.8 Modern microscopic surgery requires virtual reality and micromachine technology.

lens and the part of the body being operated upon. If the gap is less than 15 cm, it is not possible to manually manipulate surgical instruments in the gap. Another limitation is the need for highly trained and skilful surgeons. Therefore, technical improvements in microsurgery have been some what restricted since stereomicroscopic surgery was first announced in 1957 (Fig. 5.8).

Micromanipulators or microrobot hands driven remotely by microactuators would improve microsurgery by eliminating the 'minimum gap' requirement needed for manual instrument manipulation. Human hand movement would be proportionally reduced and transferred to the robot hands, and restitution forces from the object would be sensed by microsensors on the robot hands and fed back to a human hand as a tactile sensation. Visual information obtained by a stereomicroscope would also transfer as a video signal to a stereoscopic television display or a head-mounted display. These techniques are generally called 'virtual reality' or 'artificial reality' (Fig. 5.9). Thus, in the near future, surgical operation could become minimally, invasive (see Table 5.6).

5.3.3 Micromanipulation technology

Micromanipulation technology is familiar in biotechnology, especially in cell handling techniques. Cell sorting, cell fusion, injection of DNA fragments into a cell, and organelle handling: these techniques are used with many kinds of microscopes and stereomicroscopes. With ordinary microscopes we can manipulate a cell, fuse cells, and inject DNA fragments or other materials into a cell with a micropipette 1 μm in external diameter. For measuring cell function, the patch clamp method for analysing the function of a single ion channel, developed by Neher and Sackmann in 1976, also uses a micropipette and micromaniplation technique (Sakmann 1992) (Fig. 5.10).

Some microelectromechanical systems have been used for micromanipulation.

Table 5.6 Future trends of the five most common operations. (Data estimated from the health-care statistics of the USA)

	1992	1997
1	General open surgery	General video-assisted surgery
2	Urogenital open surgery	Ob-gyn video-assisted surgery
3	Open heart surgery	General surgery
4	Urological endoscopic surgery	Urogenital video-assisted surgery
5	Ob-gyn open surgery	Ophthalmic surgery

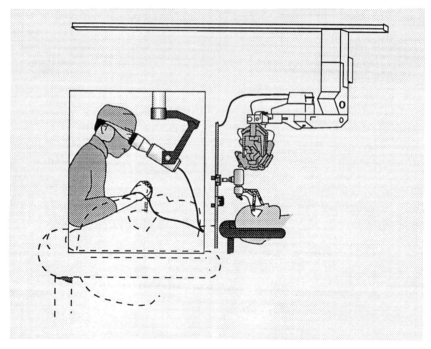

Fig. 5.9 Ophthalmological microoperation using virtual reality and microrobotic techniques. The image of the eye is enlarged in the virtual reality domain.

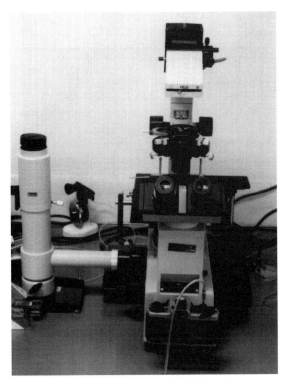

Fig. 5.10 Micromanipulation under a microscope is an essential technique in biotechnology. This method includes many of the seeds of micromachine technology.

Instead of many microscopic glass tools, especially micropipettes and micro-knives, which are conventionally made by microforge methods, moulded micropipettes are made by microelectrodischarge fabrication technology. Instead of manual cell manipulation, optical cell trapping systems have already been produced for the non-contact micromanipulation of microscopic particles, living cells, or chromosomes. Microinjection methods have also been improved by nanoprecise positioning techniques actuated by micropiezoelectric elements (Higuchi *et al.* 1987). Micromanipulation using electrostatic forces has been applied in an integrated microfluidic circuit made by a lithographic technique; this system was designed for cell fusion and chromosome handling (Washizu 1992). All these products open up on prospect of a desktop biotechnological factory using micromachining technology.

Measuring the physical parameters of living materials with atomic precision can be achieved using many types of scanning probe microscopes and also near-field microscopes. These instrument will be used to unveil many secrets of cell structure and function. Using such technology, ultramicrosurgical techniques will open up new domains such as remote microoperation, organ operation, and molecular treatment in a cell.

5.3.4 Implantable artificial organs

An artificial organ is one type of cybernetic system which includes many measurement and control functions. Most organs can today be replaced by artificial versions. Recently the member of organ donors has decreased. Therefore the development of artificial organs has acquired a new urgency (see Table 5.7).

Pacemakers   The finest and most successful artificial internal organ may well be the cardiac pacemaker, which is fabricated as a microelectronic system. Practical clinical applications started in 1960. Some types of modern pacemakers have a signal-detecting electrode to detect the sinus potential through an atrial catheter electrode, and an acceleration transducer which detects the exercise load of the patient. These signals are processed by built-in microprocessors which are designed to have particularly low electric power consumption. Lithium batteries are used for the energy source and last for ten to twenty years. Each component has been specially developed for the pacemaker, and today the pacemaker is one of the most miniaturized, reliable, and durable instruments in clinical medicine.

Sensory prostheses   Similar examples are found in sensory prosthesis, such as an artificial ear, a functional electric muscle stimulation system, and neural prostheses. These products are mainly composed of microelectronic circuits and have been made as small as possible by micromachinning.

The most popular sensory prosthesis is an artificial middle ear, which is a micromechanical device to transmit sonic vibration. Sound is converted into electrical signals and transmitted through the skin and then converted vibration into with an internal device and transferred to the stapes. This is a miniaturized hearing aid and has been in clinical use since the 1980s.

The artificial internal ear also called an artificial cochlea or cochlea implant can supply sonic signals direct to the auditory nerve. The system consists of 22-channel microelectrodes in a fine electronic bundle catheter, which is placed on the cochlea nerve. The sonic signal is converted into sonic envelope signals and divided into 22-channel signals by an external control unit. Then the signals and the electric power are sent by transformer-coupled ring antennas into a controller in the inner ear. The signals stimulate the auditory nerve. The prosthesis is a complete tether-free system and can be applied to almost all neurogenic hearing difficulties. Clinical devices were first developed and used in Melbourne University in 1978 (Fig. 5.11).

Functional electrical stimulation (FES) and neural interfaces   A functional electrical stimulation system is one kind of pacemaker applied to striated muscles. Miniaturized electrodes are placed in the muscle and electrical

Table 5.7   Incidence of organ and tissue deficiencies, or number of surgical procedures related to these deficiencies, in the United States (1991)

Indication	Procedures or patients per year ($\times$ thousand)	Indications (contents)	Procedures or patients per year ($\times$ thousand)
Skin	4750	Burns	2150
		Pressure sores	1500
		Venous stasis ulcers	500
		Diabetic ulcers	600
Neuromuscular disorders	200		
Spinal cord and nerves	40		
Bone	1 134.2	Joint replacement	558.2
		Bone graft	275
		Internal fixation	480
		Facial reconstruction	30
Cartilage	1132.1	Patella resurfacing	216
		Chondromalacia patellae	103.4
		Meniscal repair	250
		Arthritis (knee)	149.9
		Arthritis (hip)	219.3
		Fingers and small joints	179
		Osteochondritis dissecans	14.5
Tendon repair	33		
Ligament repair	90		
Blood vessels	1360	Heart	754
		Large and small vessels	606
Liver	1565	Metabolic disorders	5
		Liver cirrhosis	175
		Liver cancer	25
Pancreas (diabetes)	728		
Intestine	100		
Kidney	600		
Bladder	57.2		
Ureter	30		
Urethra	51.9		
Hernia	290		
Breast	261		
Blood transfusions	18 000		
Dental	10 000		

Sources: This table is adopted from Langer and Vacanti (1993) where it was compiled from sources that include the American Diabetes Association, American Liver Foundation, Muscular Dystrophy Association, American Red Cross, American Kidney Foundation, The Wilkerson Group, Cowen and Co., American Academy of Orthopedic Surgery, American Heart Association, National Institute of Neurological Disorders and Stroke, Source Book of Health Insurance (Health Assurance Association of America) 1991, Federal Register, and Department of Health and Human Services (Medicare-based information).

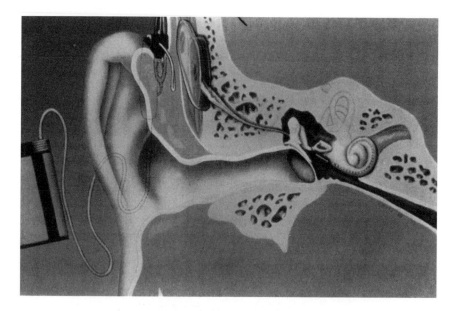

Fig. 5.11   A 22-channel microelectrode stimulator placed on the cochlear nerve acts as an artificial inner ear. This system was developed at Melbourne University.

signals from an external controller are sent by wire. The target muscle is the diaphragm for artificial respiration, skeletal muscle for artificial electrical limbs, or the smooth muscle of some internal organs such as an artificial bladder, artificial sphincter, etc. Recently, for ventricular assisting devices, skeletal muscles have been trained to contract by periodic electrical stimulation and used as heart muscles. The means of stimulation include many microcircuits and microelectrodes and require much micromachine technology.

One more important trial would be the generation of interfaces between axons and fibres of the peripheral nervous system. To simultaneously record electrical signals from many neurones and peripheral fibres, several investigators working in neurophysiology have suggested and studied the use of a sieve electrode array. A thin diaphragm with many small holes is positioned between the cut ends of a peripheral nerve and the nerve regenerates through the holes and innervates its target organ. After the nerves have innervated the target organ, long-term recording and stimulation of an axon can be successfully achieved. The record obtained in this way can help to further our understanding of the nervous system and can be used to generate control signals for manipulation of prosthetic devices for amputees and for stimulation of paralysed muscles (Edell 1986; Kovacs 1991; Kovaco *et al.* 1992).

To achieve this goal, various techniques and materials have been used to fabricate the sieve electrodes. Hollow gold cylinders, 25 μm in diameter,

Fig. 5.12 Micromachined silicon sieve electrodes have been developed and fabricated to record from the sensory nervous system and to stimulate the motor nervous system by utilizing the principle of nerve regeneration. (Photographs adapted from Akin *et al.* (1994).)

embedded in porous Teflon (Marks 1969), Teflon-coated silver wires, 77 μm in diameter, embedded in 100 μm diameter holes of epoxy resin (Mannerd *et al.* 1974), silicon substrates drilled with holes of 8 to 100 μm diameter by laser (Rosen *et al.* 1987), and tubular electrode arrays 10–15 μm high, 10–15 μm wide, and 300–1200 μm long, using stacked planar arrays of conductive belts on flexible substrates (Leob *et al.* 1977) have been reported. But these techniques have a number of shortcomings: they are complex, hand-made, lack reproducibility, have a limits to their minimum size, and are not very compatible with CMOS circuitry. Micromachining techniques are certain to improve fabrication of sieve electrodes. So far micromachined silicon sieve electrodes have been developed and fabricated to record from the sensory nervous system and to stimulate the motor nervous system by utilizing the principle of nerve regeneration (Akin *et al.* 1994). The same conceptual development has been tried in the framework of the EC ESPRIT III project INTER (Intelligent Neural in TERface), in which micromachined neural interfaces have been used to control motor and sensory limb prostheses for amputees and to directly stimulate limbs in cases of spinal cord injury (Dario and Cocco 1993).

The electrode developed by Akin at the University of Michigan (Akin *et al.* 1994) was a typical micromachined device and consisted of a 15 µm thick silicon support rim, a 4 mm thick diaphragm containing different sized holes to allow nerve regeneration, thin-film iridium recording and stimulating sites, and an integrated silicon ribbon cable (Fig. 5.12). All the microparts were fabricated using boron etch-stop and silicon micromachining techniques. The thin diaphragm was patterned using reactive ion etching to obtain different sized holes with diameters as small as 1 µm and spacing from centre to centre as small as 10 µm. The holes were surrounded by 100–200 µm² anodized iridium oxide sites, which can be used for both recording and stimulation. These electrodes had an impedance of less than 100 kΩ at 1 kHz and charge delivery capacities in the 4–6 mC cm^{-2} range. The fabrication process was single-sided and used five masks and the device was compatible with integrated multilead silicon ribbon cables. Akin and co-workers implanted the device between the cut ends of the glossopharyngeal nerve of rats. They reported that the axons functionally regenerated through the holes, responding to chemical, mechanical, and thermal stimuli.

Cortical nerve recording and stimulation   The central nervous system (CNS) includes very complicated networks. One of the most widely used techniques for studying the CNS at the cellular level is the recording of extracellular biopotentials generated electrochemically within individual neurones. A neuronal membrane depolarizes and causes ionic currents to flow in its extracellular environment after it receives enough stimuli from other neurones. The depolarized signal (an action potential) can be defected if an appropriate probe is inserted in the area surrounding the neurone, because voltage drops are associated with the extracellular current. A typical extracellular single-unit action potential is a spike with an amplitude of about 50–500 mV, with a frequency of 100 Hz to about 6 kHz. To record these action potentials, investigators insert a fine needle-like probe into the intracellular space near the active neurone without irreversible cell damage. Therefore, the size of a recording probe, which should be at its largest comparable to that of the neurone, is 50 µm or less in diameter. To get as near to the target cell as possible, which means detecting the signals with the highest signal-to-noise ratio, the electrode should be as small as possible.

Neurophysiologists have used three types of microelectrodes: metal, glass, and photolithographically processed. Metal electrodes are made of platinum, gold, stainless steel, tungsten, or molybdenum, processed by etching, and the diameter at the tip is less than 0.1 μm (Skrzypek and Keller 1975). Metal microelectrodes are most suitable for extracellular recording of AC signals. Glass micropipettte electrodes are made from a 1–2 mm glass tube. The tube is pulled under heating until the diameter of the tip is between 1 and 0.1 μm. The pipette is filled with an electrolyte and can be used to measure DC and low-frequency biopotentials.

Recently neurophysiologists have begun using multimicroelectrodes to analyse complex networks of the central nervous system. Multielectrode probes are needed for long-term recording and stimulation. In clinical applications long-term connections with neurones for delivering control signals to prostheses become important. These multielectrodes should also be capable of incorporating electronic circuitry. As a first stage, passive microelectrodes have been used in neurophysiological research on the central nervous system, and have answered many of our questions about the enormous complexity of neural structure.

A passive microelectrode is defined as one which does not contain any interfacing electronic circuitry on the electrode substrate. Photoengraved microelectrodes are used for the multimicroelectrode and are fabricated using the silicon process. They are fabricated by depositing and patterning thin-film electrodes on a thick substrate. Substrate materials used for them have included silicon, tungsten, molybdenum, glass, polyimide, and other insulators. The thin films used as dielectrics have included polyimide, silicon oxide, silicon nitride, PMMA, and glass, and the electrode conductor materials have included gold, platinum, tungsten, tantalum, and nickel. Using such materials, the first micromachined electrode was developed in 1970 by Wise, who fabricated gold recording electrode arrays on a silicon substrate.

A typical micromachined passive multimicroelectrode was reported by Najafi *et al.* (1985). A single-shank electrode with multichannel recording sites was first developed, followed by probes with multiple-shank electrodes and a three-dimensional multielectrode (Hoogerwerf and Wise 1991). A three-dimensional recording system, comprising a 32-electrode active multiple-shank probe precisely positioned in a micromachined silicon platform, has been implanted in the visual cortex of guinea pigs. Up to four months after implantation, the arrays of the system showed no significant signs of damage, and detailed recordings have been made using passive systems. A typical shank is 1.5–3 mm long, 30 μm wide, and 15 μm thick. These probes are strong yet flexible, and can penetrate an intact pia-arachnoid membrane with no noticeable bending and minimal dimpling of the cortical surface (Najafi and Hetke 1990) (Fig. 5.13). The micromachining procedures was as described below.

The supporting substrate was made of a silicon wafer of (100) orientation. The wafer was first thermally oxidized and the oxide removed selectively using photolithography to expose silicon areas in the shape of the intended pad. The wafer was then subjected to deep boron diffusion (to a depth of about 15 μm) using the thick oxide area as a selective mask. Following boron diffusion, the wafer was again thermally oxidized and additional dielectrics (silicon nitride)

 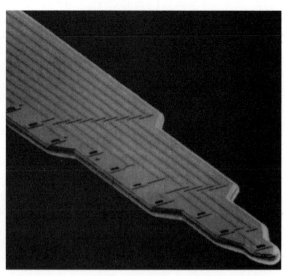

Fig. 5.13   This three-dimensional recording system, which is composed of 32-electrode active multishank probes precisely positioned in a micromachined silicon platform, have been implanted in the visual cortex of guinea pigs. (Photographs adapted from Najafi and Hetke (1990).)

deposited on the wafer using chemical vapour deposition (CVD); this served as the insulating layer. To pattern and etch the top dielectric films, polysilicon or refractory metals such as tantalum, tungsten, and iridium were coated with silicon dioxide by a CVD process. After deposition, the polysilicon or refractory metal was patterned using photolithography and reactive ion etching (RIE). By putting another resist on the insulated recording sites and the output bonding areas, openings were etched through the upper dielectrics to expose the interconnect material. These surfaces were etched free of any residual oxide and coated with successive layers of chromium–gold or titanium–iridium. The resist was then removed by a lift-off process, removing the overlying metal from the wafer everywhere except in the recording site and bounding pad areas. Finally, the wafer was thinned using a back-side unmasked etch in a hydrofluoric–nitric acid mixture. The front of the wafer was protected with wax during this thinning. Next, the wafer was placed in an anisotropic silicon etch composed of ethylenediamine, pyrocatechol, and water at a temperature of 115 °C.

Fabrication and development of a multichannel probe including a signal processing circuit on the same substrate as the recording electrode has been made possible only recently by micromachine technology. Such a probe is called an active probe. Active probes become important for prosthetic applications, where the probes are interfaced to microcomputers for processing of control signals. The merits of active probes include decreasing and minimizing the number of output leads and bonds, and thus the effects of leakage from the output, amplifying the recorded signals, changing the format of the output data, providing an on-chip stimulation capability, allowing for remote testing, and adding programming capability to the probe.

The first attempt to fabricate active microelectrode arrays was made by Wise in 1975 and extended by May, White, and co-workers in 1979 to develop stimulating arrays for an auditory prosthesis. However, those circuits were made with hybrid connection of the electronics to the sensor and were not true monolithic chips. In 1984, Takahashi and Matsuo became the first to fabricate a multielectrode silicon probe with interface circuitry for amplification and multiplexing using planar and three-dimensional fabrication techniques, but they didn't test any practical applications. The first practical products were reported by Najafi and Wise in 1986. The on-chip electronics consists of preamplifiers which amplify each signal from multiple channel electrodes followed by an analog multiplexer and a broad-band output buffer to drive the external data line, all powered by a battery. The implantable circuitry requires only three output leads; for power, ground, and data. In order to allow external regulation of the on-chip sample clock, a synchronization pulse is inserted once each frame as an eleventh channel. The external electronics strips off these pulses, generates the sample clock, and demultiplexes neural signals for external recording/processing equipment (Ji *et al.* 1991). An active multielectrode shank is 15 μm thick, contains ten gold recording sites with 100 μm centre-to-centre spacing, each with an area of 64 μm^2, and has a 3.2 mm shank that tapers from 10 μm close to the tip to 160 μm at the bottom. The circuitry occupies an area of 1.3 mm^2 and consumes about 5 mW of power from a single 5 V power supply. A 32-electrode active recording microprobe has also been developed (Ji 1990).

Many of these applications requires the transfer of data across the skin. These techniques also depend upon micromachining. Silicon-based multilead ribbon cables are flexible, strong, and can support multiple interconnect leads in a small volume (Hetke *et al.* 1991, 1994). A telemetry technique for transdermal data and power transfer has also been developed in a miniature implantable system (Akin *et al.* 1990). For practical clinical application as neural prostheses, a number of electrically interconnected, self-contained recording and stimulating units should be incorporated. Although stimulating electrodes and systems have already been developed for auditory prostheses, similar systems have not yet been successfully employed in visual prostheses. In order to construct such prostheses we must develop high-performance hermetic packaging (Ziaie *et al.* 1993) and implantable power sources that occupy a small volume and have a high energy density.

Heart valve prostheses   Heart valve prostheses are one of the best developed artificial organs. They are not strictly micromachines but give rise to many of the technological problems that are encountered when micomachines are implanted in the body. Since the first clinical application of a ball valve was announced in 1961, many kinds of heart valve prostheses have been developed and used. The valves are classified as mechanical or biological valves according to the materials used. Recent mechanical valves are made of pyrolytic carbon and consist of tilting disks. Biological valves are made from the aortic valves of pigs processed in glutaraldehyde. As valves must function in the cardiovascular system for more than five years, they must be very durable, have good blood and tissue compatibility, and maintain adequate valve function for a very long period. These requirements are fundamental for all implanted artificial organs.

Ordinary valve prostheses are more than 10 mm in diameter, but for newborn babies or for some implantable artificial organs valves are required to be less than 3 mm in diameter. A jellyfish valve is one type of new mechanical valve prostheses under development. It consists of a centrally fixed leaflet valve made of polyurethane fabricated by moulding (Imachi *et al.* 1993). Many microvalves will be required to handle the circulation of blood in other artificial organs.

Vascular prostheses and internal shunts   The first vascular prosthesis was used as a homograft. In 1954, the first clinical case using a processed polymer graft was reported. Today, vascular grafts are one of the most successful artificial organs in clinical medicine. The inner diameter of clinically applicable vascular grafts must be greater than 3 mm because grafts less than 3 mm in diameter cause severe blood coagulation. Bad blood compatibility in fine-diameter grafts is due to the structure of artificial vascular grafts. The structures of modern vascular grafts are classified as woven, expanded, or smooth. Polyethyleneterephthalate (Dacron) is used for woven grafts, polytetrafluoroethylene (PTFE) is expanded grafts (expanded PTFE (EPTFE), Goretex), and many blood-compatible polymers have been tested for smooth surface grafts with a small internal diameter. Woven grafts are used as larger-diameter prostheses for aortic graft applications and EPTFE is usually used for grafts of 3–7 mm in diameter in limb arteries and the inner shunts for dialysing. But these grafts cannot be used for chronic implants requiring orifices less than 3 mm in diameter. The diameter of most arteries of important organs, peripheral arteries, and conduits of artificial organs is less than 3 mm. We have many demands for small autithrombogenic grafts. Culturing endothelial cells on the inner surface of the graft and coating with antithrombogenic materials such as adhesive protein, surfactant materials, block copolymers, etc., is now being investigated. These technologies would become essential for the development of micromachines for medical use and belong to the field of artificially structured materials (see section 9.2).

Artificial skin and access plugs   The technology of artificial skin also belongs to that of artificial structural materials. Artificial skin for temporary use is made from biological materials (pig skin, collagen, chitin, etc.) and processed polymers (woven nylon coated with collagen, polyurethane membrane, polyleucine membrane, etc.) and applied clinically for preventing infection and body fluid loss, for epithelialization (to cover epithelial tissue on the wounds), for pain relief, and for protection from injury. Permanent artificial skin acts as a material supporting the organization of cutaneous tissue; it is composed of a matrix of polyglycolic acid or collagen.

An access plug is an artificial part of the skin which acts a transportation gate or conduit across the skin barrier. A clinical access plug consists of a hydroxyapatite base supporting an injection port or signal cables. As hydroxyapatite has good tissue compatibility, drugs and nutrition or signals can be transported through the access plug. An access plug is sometimes called a skin button and is an essential part of implantable artificial organs and micromachines.

Lenses   Contact lenses are used by some five to seven per cent of people in developed countries. As a contact lens is worn directly on the cornea, the material from which it is made must have tissue compatibility, must be permeable to oxygen, must withstand repeated insertion, removal, and sterilization, and must have good optical characteristics. Hard contact lenses (HCL) are made of polymethylmethacrylate (PMMA), siloxane, or its fluoride composites. Soft contact lenses (SCL) are made of polyhydroxyethylmethacrylate (PHEMA) processed to a water-containing gel.

Contact lens are good platforms for implanting micromachine devices. Some investigators are developing sensors and transducers in contact lenses to measure physiological data non-invasively.

Intraorbital lenses (IOL) are used for treatment of cataracts. They are made of PMMA or silicon. In the USA more than 15 per cent of people over 80 use IOL (some 500 000 cases annually). The functions have gradually been improved from one-piece lenses to three-pieces lenses and from fixed focus to multifocus lenses. IOL offer a good field of application for micromachine devices.

Other major artificial internal organs   Other large internal organs or limb prostheses are not easily miniaturized. One reason for this is that the necessary mechanical parts are too large; there is a great incentive therefore to develop micromechanical parts for such artificial organs.

The most useful micromechanical parts for such applications would be microactuators and microenergy conversion technology. Live organs use chemical energy in the body; in man-made systems, energy sources must be supplied externally. If we manufacture tether-free internal organs, we must develop a system which can transmit through the skin or a high-density energy reservoir. Energy can be supplied from outside the body in the form of electromagnetic force, magnetic force, or ultrasonic force. Recent development of microelectrostatic motors and ultrasonic motors will be a key technology in this field. But the most important products will be microscopic linear actuator elements which can be integrated like sarcomeres in muscle fiber. A film-type actuator powered by the

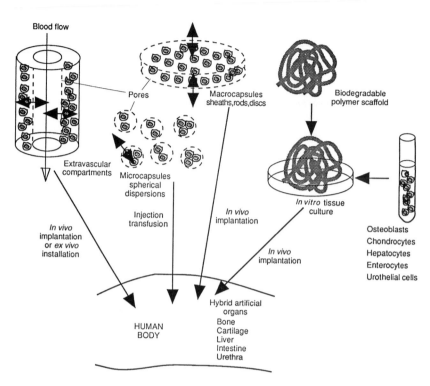

Fig. 5.14   There are three common closed-system and one open-system configuration for cell transplant devices that are usually called hybrid artificial organs. In vascular devices, the cells are placed in an extracellular compartment surrounding a tubular membrane (inner diameter ~ 1 mm) through which blood can flow. In macrocapsule systems, the cells are placed in sheaths, roads, or discs (diameter ≥ 0.5 to 1.0 mm). Macrocapsules and vascular devices often consist of acrylonitrile–vinyl chloride copolymers. In microcapsule systems, the cells are placed in injectable spherical beads (diameter < 0.5 mm). Microcapsules are commonly made of hydrogels. In one approach to open system implants, three-dimensional highly porous scaffolds composed of synthetic polymers serve as cell transport devices. The polymer could be degradable or non-degradable. Materials that disappear from the body after they perform their function obviate concerns about long-term biocompatibility. (Figure was adapted from Lanza *et al.* (1992) and Langer (1993) and combined and modified.)

linear electrostatic principle has already been developed and used (Egawa *et al.* 1991). It may be possible to produce an artificial muscle system using an electrostatic linear actuator. A linear microactuator element which is able to use vibration energy has been reported, and it is possible to integrate this element to make a muscle-like structure (Fujimasa 1992).

For developing artificial endocrine organs such as the pancreas, adrenal organ, etc., we must prepare some macro drug delivery systems (Fig. 5.14). An artificial

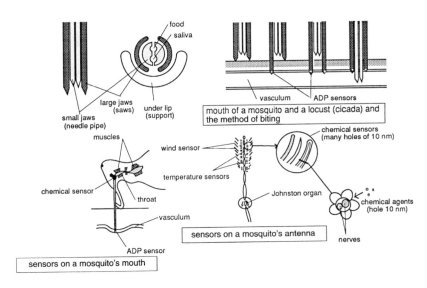

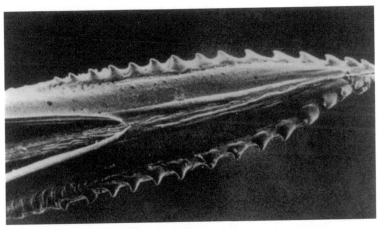

Fig. 5.15 The mosquito has a pair of microsaws as part of its blood-sucking mouthparts. At the tip of the saw we can see the orifice of an ADP (adenosine diphosphate) sensor inlet. (From Niillson (1976).)

micro endocrine organ, for example an artificial pancreas or islets of Langerhans, requires an ultramicroscopic glucose sensor, drug container, and drug releasing system. Alternatively one could use a micro injection system resembling the mouth parts of a mosquito (Fig. 5.15). Recently an injection system has been developed which has a fine needle with serrations (Komine 1994).

A microsecretory gland which looks like a pill and is used for measurement and treatment in the intestinal tract is usually called an endoradiosonde. A prototype was first reported 30 years ago, but recently it has been revived as the first proposal for a medical micromachine proposed by the Ministry of International Trade and Industry of Japan (MITI) in 1991.

Artificial oxygen carriers (artificial red blood cells) are also micromachines. There are three current approaches to their development. The most successful approach is the development of oxygen-carrying fluids which are inert in the human body. Perfluorocarbon is a stable oxygen carrier which is almost inert in the body; it has been used as a blood substitute in PTCA operations. Microencapsulation of haemoglobin obtained from other animals is another method of making artifical red blood cells. Recitine-encapsulated haemoglobin has been developed and tested. Covering and hiding the immunological branch of haem protein is another method of producing artificial red blood cells. The immunologically important part of the haem protein was covered by glutaraldehyde, but this method has not yet been applied clinically.

Drug delivery systems

Drug delivery systems using microcapsule technology are another application for micromachines in medicine. Today's drug delivery microcapsule technology is mainly dependent upon constructing liposome-coated drugs to be applied in vessels. Two important technologies are involved: how to deliver the drug to the target organs or ,issues at the most beneficial time and how to escape from phagocytes. We must develop capsules which can be destroyed by an external signal and methods of detecting capsules from outside the body. In one trial microcapsules have been made from some kind of polymer film which can be detected and destroyed by external ultrasonic power excitation (Ishihara et al. 1991). But this method cannot deliver drugs outside of vessels or into cells; some functions must be added to enable drugs release outside of vessels which would require machine parts of submicrometer size.

Micromechanical systems technology will have a great impact on biomedical engineering. Technologies for the creation of submicrometer sensor elements would greatly advance the development of micromachines in medicine. Recent surveys of microsensors have reported that the electronic transducer is a barrier to miniaturization below 10 μm. We must find some other principles for detecting chemical or physical parameters in the living body. Optoelectronics would supply us some methodologies, such as photon technology and fiber optics. Remote sensing technologies can also give us much information about the inside of the body from the outside. Automated multichannel biochemical sample measurement systems (autoanalysers) have advanced greatly over the past 40 years since the system components have become miniaturized; the whole system can now be built on a desktop. Today, new types of machines for sorting cells, organelles, and DNA fragments are being developed as automatic desktop machines. Microscopic measuring transducers are required for such machines. Recently, multiplex gear sensing chips were developed. Electronic hybridization of a DNA probe on a very large scale integration (VLSI) chip was reported by Nanogen Inc. and the 'Genosensor consortium' (Heller 1996). Biochip array technology represents a new and exciting micromachine development.

There are no existing technologies which meet all these requirements. Even those which have potential are still at the experimental stage, in the development stage, or just on the drawing board; many of them will be important basic technologies in the near future.

6
The future for micromachines

I have described a future picture of micromachine technology based on the present technical developments. Many new technologies will appear with increasing speed in years to come. However, the technology is still not applicable to biomedical uses. Why?

The reason might be the absence of technologies for the production and design of micromachines. Almost all modern micromachine fabrication techniques are derived from other industrial technologies. The silicon process is the most important technique for micromachine fabrication, but the machines produced are thin and flat and use silicon derivatives. Their energy conversion mechanism uses ordinary electronic and magnetic principles. Have any original techniques been developed for micromachine fabrication? Many proposals have already been discussed in this book, but the most useful technology might be obtained by biomimetic analysis of self-assembly and energy conversion in living organisms.

When we observe living things as machines, we become aware that the autonomous machine, which we usually call life, does not exist at a size of less than 1 μm. We know that the size of proteins is from a few to a few hundred nanometers. If we can use these parts and construct a machine of 1 μm^3, the number of parts may range from 10^3 to 10^6. Why are there no autonomic machines in the 1 μm to 100 nm size range? Certainly the size of viruses is less than 1 μm, but they do not possess their own energy conversion systems. At what stage in evolution did life develop an energy conversion system?

According to my speculation, a size of 1 μm is fundamental for life. It is well known that a similar barrier exists governing the size of microelectronic elements and devices. When an electronic element like a transistor becomes smaller than 1 μm the element does not work in the usual way. We should take the quantum effect into consideration, and it is for this reason that quantum elements appeared in electronics. In order to analyze life as a machine, we must take the effect of molecular motion at body temperature into account. We cannot ignore Brownian motion when dealing with sizes of less than 1 μm. Some new energy conversion mechanism may be beyond the 1 μm barrier. When we want to make truly micrometer-sized machines, we must observe the 'machine of life' and imitate the principle of its energy conversion system.

The domain between 1 μm and 1 nm includes many supermolecules that interact strongly with each other. This domain has recently been called the 'mesoscopic domain'. Many measurement tools and fabrication techniques have been developed and proposed for construction of future micromachines. Many investigators of nanotechnology, however, have worked up from atoms and molecules. Their methodology depends upon analysis extrapolated from the molecule to the system, 'bottom-up' style. Protein engineering is the most powerful tool for analysing and producing mesoscopic structures. However, I have tried to analyse and fabricate mesoscopic machines from system to molecule, using 'top-down' methodology. I will describe such technology in Part II.

Part II
The mesoscopic domain of micromachines

Introduction

Humans are defined as animals who make machines. Human beings must have made many machines without clear physical principles for the design, but we have never developed micromachines smaller than 1 mm. In a wrist-watch, an example of one of our smallest machines, the smallest parts are not less than 1 mm. If micromachines are limited to 1 mm, the size of microparts is inevitably less than 1–10 μm. What factors act as a barrier for producing micromachines (Table II.1) and what kinds of differences exist in the micromechanisms of living organisms, which are able to construct mechanical parts of less than 1 μm (Table II.2)? Are there any hidden physical principles or powers to actuate such micromechanisms? Can we discover and artificially apply such physical principles and powers in engineering? Mesoscopic engineering is aiming at finding such physical principles and producing micromachines in the micro- and nanometer domain.

Recently, many observation techniques have been developed for the submicrometer domain (Table II. 3). Such techniques include many kinds of microscopes such as near field scanning microscopes, and scanning probe microscopes, laser tweezers, and microfibre chemical sensors, with which we can measure many kinds of physical and chemical parameters of cells and living micro- and nanostructures. Many hidden principles of cell structure and function are being unveiled. We have begun to obtain general principles from such cell functions which can be applied in the design of microscopic actuators and robots. It has already been mentioned in Part I that there has been great progress in microfabrication technology in industry. Many microprocesses, such as the silicon process and lithography technology, microelectrodischarge milling, optical resin moulding, the LIGA process, etc., have been used in micro- and submicrofabrication. Today, some micromechanical parts have become commercially available.

Table II.1 Micromachine characteristics in classical Newtonian mechanics

---

Relatively small force of inertia and large force of viscosity
Surface effects become dominant
Mobile machines are dominant
Heating and cooling of an element is easy but thermal insulation is difficult
Heat dissipation of a highly integrated actuator is difficult
Fluidic machines become beneficial

Using such a technological background, we can just begin to apply mesoscopic engineering in industry. The precursor to mesoscopic mechanical engineering is analysis of cell structure.

Table II.2 Differences between cell machines and human-made micromachines: starting points of mesoscopic mechanical engineering

Size: submicrometer
Weak bond force of supermolecule and highly integrated
Actuation energy: almost the same level of thermal molecular noise
Uncertainty relation between input energy and output work:
    chaos, fluctuation, and non-equilibrium
High efficiency at ambient temperature and at atmospheric pressure
Autonomous and disseminated actuator system
Self-assembling and self-supporting

Table II.3 Observation and analysis techniques of organelles as mechanical parts

Techniques	Representative instruments
Microscopic observation	Electron microscope
	Laser confocal microscope
Scrap and collection	Column chromatography
	Electrophoresis
Read blueprints	Genetic engineering techniques
Structure analysis	X-ray diffraction
	MRI
Dynamics observation	Fluoromicroscope
	Laser confocal microscope
Self-assembling	Micropipetting of gene fragment
In vitro testing	Laser tweezer
	Scanning probe microscope
	Laser confocal microscope
Chemical analysis	Submicrometer fiber sensor

7
Invisible machines

7.1 Mechanics in the mesoscopic domain

What role can nanotechnology play in the development of micromechanical devices? In this chapter we will find out what a large impact nanotechnology can make upon the development of micromachines. Machines fabricated and processed using nanotechnology will obey completely different design rules and be made of unusual materials. The master model would seem to exist in the form of tissue and cell structure.

We usually observe and measure the dynamic movement of microtargets under an optical microscope. When the size of the targets is less than 1 μm, we cannot observe them in the living state and we cannot measure their dynamic functions with any conventional optical microscopes or measuring equipment. Electron microscopes can be used to observe nanosize structures, but such structures must be prepared by drying or freezing in a vacuum. X-ray diffraction has been used to study dynamic nanosize organelle movement, but the data are analysed in batch mode. Therefore, nanometer organelles can usually be observed only in a steady state. We then estimate the dynamic functions with simulation models. In spite of the fact that all the mechanical principles of cells work on a nanometer scale, we have designed our micromechanical systems without observing the way in which cellular mechanisms work. This is one of the main reasons why our micromachines cannot achieve the efficiency of an organelle.

From biological systems we can discover many new concepts for use in the design of mesoscopic machines, and the basic physical principles which obtain in the mesoscopic field. In the new era, many supporting technologies will appear such as: new membrane fabrication techniques, new autoassembly theories using chaotic dynamics and fractal theory, ultra fine supramolecules and particles for new materials, such as many kinds of fullerenes, etc.: these technologies will be the new tools for producing mesoscopic machines, and will open up a completely new world of machines.

The size of the fundamental parts of nanomachines will range from 10 nm to 1 μm. From 10 nm to 1 mm, we have a multiplication gap of 10^5. When we think of such a range in terms of modern industrial mechanical parts, the size of the finest nuts in a wrist-watch is almost 1 mm and the size of a long rail of the Japanese bullet train or that of a large oil tanker exceeds 100 m: between 1 mm and 100 m is the same multiplication gap of 10^5. This wide size range

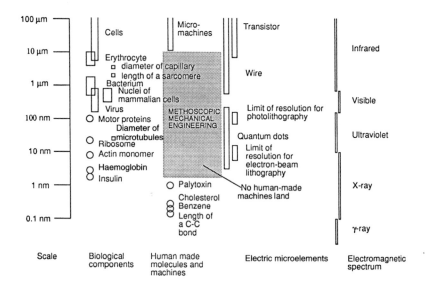

Fig. 7.1 Sizes of machines made by living systems and humans. We have made hard machines larger than 10 μm. We can produce polymers or some molecules by assembling until 1 nm. From 10 μm to 1 nm, we have not produced any micro-mechanical parts. Moreover, there are no actuators to convert energy to mechanical work. In such a domain we must find a new principle of dynamics to design new actuators. We call mechanical engineering in this domain 'mesoscopic mechanical engineering'.

encompasses almost all man-made mechanical parts. The micromachine world therefore will include almost the same number of machine parts when we are able to develop suitable micro- and nanofabrication technologies.

Electronic products already range from centimeter to submicrometer size. An integrated circuit includes many submicrometer transistor elements and micrometer microprocessor circuits, and these microelements invite unlimited miniaturization of electronic devices. As examples, micro-CPUs or charge-coupled devices (CCD) which include more than 100 000 elements have been developed and applied in home video cameras or personal computers (see Fig. 7.1).

If we can develop techniques for the fabrication of submicrometer machine parts and add some molecular handling techniques like protein engineering which make it possible to make molecular machines, the technology would become a true micromachining technology; it would then be possible to make small movable machines or robots comparable in size with recent electronic microdevices. In the future it may be possible to invent an assembly process which can compete with that of biological systems. Such micro- and nanotechnologies might exceed the possibilities of current biotechnology and have wide industrial applications.

7.2 Cells as machines

7.2.1 Mechanical parts of cells

The size of the fundamental moving parts in a cell is usually less than 1 mm; such parts are called 'organelles'. The size of a cell, which is a component of organs and tissues, is usually 10 μm or more. The smallest transportation conduits in the body are capillaries and their external diameter is almost 5 μm. The smallest living organisms are bacteria, around 1 μm in size. There are no independent living organisms smaller than 1 μm. Eukaryotic cells have a complicated structure, and the size of the main cell structures such as cell skeletons, motor proteins, and organelles is less than 1 μm. Using these submicrometer components, the living body constructs fine and intelligent machines (see Fig. 7.1).

Biological machines, which are simulated biological systems, have until now relied on mimicking the functions of living systems, and the engineering using such concepts is called 'biomimetics'. However, micro- and nanomachines simulate the structures of biological systems, and we intend to use their mechanical parts to make not only miniaturized machines but also large machines like animals and plants. This new field has been called the science of microbiomechanisms; it is influenced by cell anatomy and is working towards a technology for designing mesoscopic machines. Microbiomechanisms in the living body are powered by biological actuators (Fig. 7.2) whose functioning we hope to duplicate in nanomachines.

7.2.2 Flagellar motor

Bacteria are prokaryotic cells. The bacterial cell membrane encloses mainly fluid and soluble biopolymers. Bacteria more using chemically powered motors (flagellar motors), which are located on the cell membrane. A flagellar motor is made of a few types of proteins and their genetic codes and self assembly sequences have already been analysed (Fig. 7.3). The mechanical functioning of flagellar motors has been observed under a scanning laser microscope. The 10 μm flagella rotate at 15 000 rpm (Kami-ike *et al.* 1991). The energy source is thought to be the proton gradient between the cytoplasm and the external environment, but the chemomechanical energy conversion mechanism is still unknown. The flagellar shaft might be supported by a bearing which is suspended by van der Waals atomic forces. In spite of the fact that the flagellar motor rotates at such high speed, it is said that the motor and flagella never break nor wear out. The speed at the edge of a rotating flagellum is very low, because the flagellar diameter

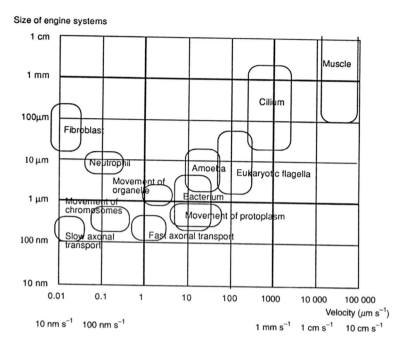

Fig. 7.2 Moving sreeds of cell and organelles. When we observe living organisms, especially cells and organelles, they have some kind of engines. The size of the engine relates to the size of the living system. Large animals have high-velocity engines such as muscles and small unicellular systems have low-velocity engines such as cilia or eukaryotic flagella. Those engines consist of motor protein and cytoskeleton complexes which are highly integrated in the cells.

is only 100 nm. The movement therefore produces only a very weak impulse power on the nanostructure (see Fig. 7.3).

7.2.3 Motor proteins and cell skeletons

Eukaryotic cells have skeletal networks. The main components of the skeleton are intermediate fibers, microtubules, and actin chains. The cells rely on skeleton-based protein motors to generate intracellular movements. The skeleton is basically composed of protein monomers such as actin and α and β tubulin. It acts not only as a structural support but also as a corridor for organelle transport. The skeleton is usually polarized and the polarity produces the direction for transportation.

Microtubules, actin chains, and other structures are connected by (or bridged with) short protein monomers. The bridge protein length is about 50 to 100 nm.

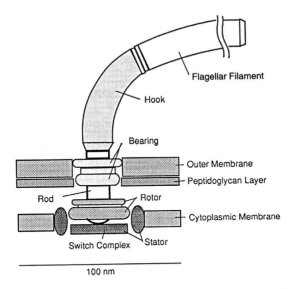

Fig. 7.3   Structure of flagellar motor. The motor rotates at 15 000 rpm using the proton gradient. However, the energy conversion mechanism is still unknown.

Thus, nanometer-scale network structures are constructed in a cell. Recently it has become clear that the short protein bridges mainly consist of motor proteins or proteins of similar structure. Dyneins and kinesins are typical proteins found in microtubule motors. These motor proteins have been observed to move along the microtubules and to form bridges between microtubules and other cell structures. They usually have two heavy-chain heads which include ATP-ase to provide energy for moving (Goldstein and Vale 1991).

To our surprise, some kinds of motor protein act as construction tools for microtubules or other parts of the cytoskeleton (Hirokawa 1991). In a neuronal axon it was observed that a tail of kinesin motor attaches to an organelle and the head is attached to microtubules; thus the kinesin can transport organelles from the electrically negative end to the positive end of a microtubule using high-energy charged phosphoric acid (Hirokawa *et al.* 1989).

These observations suggest to us how future nanotools could be used to manipulate nanoparticles and what functions are required for nanoprobes. A model of a nanoprobe system already exists in a cell. Unfortunately, we have no information about the trigger signals which combine external stimulation and formation of internal structure the concentration of Ca^{2+} ions in a vascular endothelial cell, caused by increasing cell wall shear stress, is the only chemical signal which has so far been related to cell growth (Ando *et al.* 1987).

7.2.4 Integrated actuators: the sarcomere

For animals, the most popular, most effective, and fastest actuator system is probably striated muscle. The fundamental element of striated muscle is the sarcomere, which has a length of 2.5 µm and thickness of 10–20 nm. The sarcomere is a chemomechanical engine unit which works with high efficiency (tunable from 0 to 80 per cent) in a constant-temperature environment. The switching time from stasis to constriction is relatively long, and it takes almost 100 ms.

The power and the contraction are produced by the interaction between chain-like actin cytoskeleton and myosin motor proteins. The energy source is believed to be hydrolysis of ATP. In three dimensions integrated sarcomeres show a hexagonal cylindrical crystal structure, which has a myosin bundle in the centre. Many of the actuators that make up the structural organelles common to all cells were first discovered in muscle, and muscle contraction is the most typical and the best understood of all kinds of machine-like movement of which animals are capable. Muscles are composed of repeated assemblies of thick motor protein filaments (myosin molecules) and thin cell skeleton filaments (actin molecules; Fig. 7.4). These filaments are regularly arranged and comprise an actuator unit called a sarcomere. The sliding motion is caused by molecular interactions between adjacent thick and thin filaments. It is thought that the motive power of the slide is produced by a large conformational change in the myosin motor head which is activated chemically by hydrolysis of ATP. The movement is called a 'power stroke', the idea being that the consumption of one unit of source energy produces one unit of mechanical work (Huxley 1957).

However, from a recent *in vitro* study using myosin monomers and actin filaments, it has become clear that the relation between chemical energy supply and physical work output of actin filaments is not exactly one to one. The distance moved due to conformational change in a myosin head per period of an ATP hydrolysis cycle is estimated to be 20 nm at most (because the length of a myosin head is 10 nm), but the distance moved by an actin filament of the *in vitro* study was found to be more than 100 nm (Harada *et al.* 1990). The reason is that the contraction does not occur due to a definitive energy conversion mode such as one unit molecular conformational change induced by ATP hydrolysis but happens because of some other power contribution having probabilistic mode owing to the severe thermal noise in the molecular environment. In the resting position, almost all the myosin heads become separated from actin filaments and, at contraction, the power of quantum statistical mechanics activates the myosin heads. The mechanism of muscle contraction is still unclear, but some unknown energy conversion mechanism must exist as the nanoscale actuator, which may provide us with some ideas for future nanomachine design. However, the energy conversion mechanism is still not understood, especially that required to produce contraction. Huxley's famous cross bridge model (Huxley 1957) and its many modifications have some difficulties in explaining the movement in excess of 100 nm of an actin

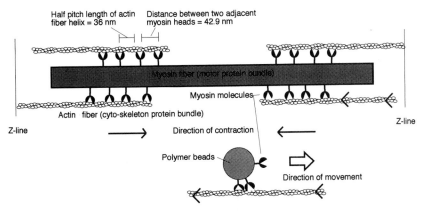

Fig. 7.4 Schema of sliding movement of a myosin fiber on actin filaments. The essential elements of striated muscle consist of myosin fiber bundled up motor proteins and actin fiber composed of cytoskeleton proteins. Contraction of the muscle is caused by myosin fibers glided along actin fibers. If we attach myosin monomers on a fine plastic bead which is also coated with a flourescent substance, and put it on an actin fiber *in vitro*, we can observe the bead move along the actin fiber when supplying ATP. The molecular structure of the heavy chain of the myosin molecule was finally unraveled in 1993. As the molecular structure of the actin monomer has been completely analysed, we expect the molecular interaction of the actin–myosin complex to be discovered in the near future.

filament within a single ATP hydrolysis cycle. This movement has been observed *in vitro* with a luminescence microscope (Ishijima *et al.* 1991). In Huxley's original model, the conformational change in a myosin head producing actin filaments movement caused by an ATP hydrolysis cycle was estimated to be only 10 nm; many models of contraction have since been proposed to explain this discrepancy. An electrostatic linear motor is a typical example of a non-contact energy conversion system which might be used to model muscle contraction (Yano *et al.* 1982; Jacobsen *et al.* 1989 (Fig. 7.5)). As a spin-off application of the principle, a microstructure linear motor called a 'multilayered electrostatic film actuator' has been developed by Egawa and Higuchi (1990).

When we analyse dynamic physical movement in the mesoscopic environment, where thermal molecular turbulence is dominant, we must consider Brownian motion. We have proposed a new motor element which is actuated by vibration energy. The structure looks like a sarcomere. The element catches field energy, as impact forces, through its asymmetrical teeth (Fig. 7.6) (Fujimasa 1992).

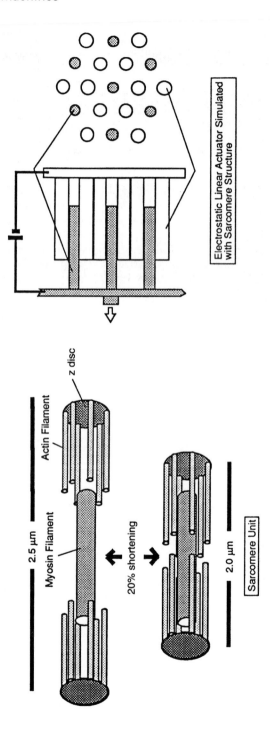

Fig. 7.5 The structure of an actin–myosin complex in a sarcomere looks like that of an electrostatic linear motor proposed by Jacobsen *et al.* (1989).

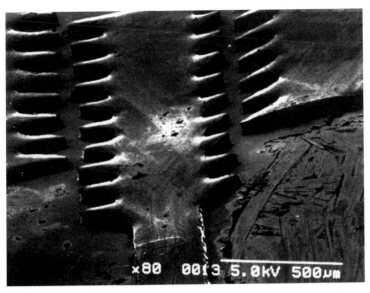

Fig. 7.6   A symmetrical teeth derive energy from field vibration and are made to collide with each other. The slider (centre) in moved by the impact of the collisions.

7.3   Observation and manipulation methodology

7.3.1 The possibility of nanomachine manufacture

Can we produce nanomachines in the near future? Using recently developed manufacturing techniques, we may produce some micromachines having a size of a few micrometers, because we can inject some genes and materials into a cell using micropipetting techniques. But we cannot make nanomechanical devices 10 nm long. If we were able to develop some tools to produce such nanoparts, and we could assemble the parts to make a machine, the prospects for manufacturing nanomachines would be excellent. Using recent protein engineering we can produce some proteins. The protein molecule produced is a functional mechanical part in the nanometer domain, but the manufacturing 'factory' is still a living cell, which is a typical black box for mechanical engineers.

Are there any production techniques with which we can pick up and hold a molecule or an atom and put it in a certain position? If we had such an atomically precise technique we might be able to make completely new materials such as new membranes with ionchannels, artificial liposomes with flagella, and so on. Such technology was thought impossible until 1984. But rapid progress in subnanometer electronic positioning methods and the fabrication techniques of

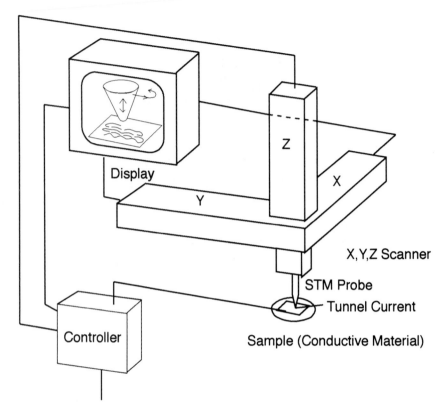

Fig. 7.7 Conceptual stucture of the scanning tunnelling microscope. The STM probe is scanned along *x*, *y*, and *z* axes with atomic precision driven by piezoelectric devices.

atomic size probes have now made it possible to measure material surfaces with atomic precision and to position and move atoms or molecules on the surface. This technique has been called scanning probe microscopy (SPM). Using SPM, we can manipulate atoms and molecules, and make an atomically precise structure.

In the near future we will have the engineering capability for nanofabrication. This is one of the reasons why the micromachine age is now downing.

7.3.2 Manipulation techniques: scanning probe microscopy (SPM)

Many kinds of scanning probe microscopes have been invented and announced after the first report of a scanning tunneling microscope by Binning and Rohrer (1982) (Fig. 7.7). Recently, Wickramasinghe (1991) mentioned more than 50

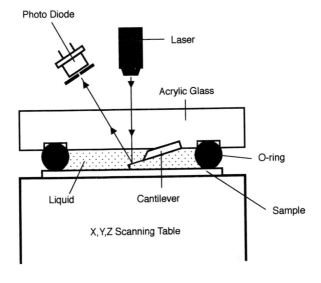

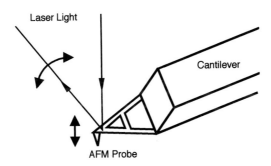

Fig. 7.8 Conceptual structure of the atomic force microscope. When we apply the probes to the surface materials, we can measure their physical characteristics and fabricate their surface (illustrated by T. Chinzei)

kinds of SPM. Many kinds of physical parameters have been observed and converted to molecular size patterns (Fig. 7.8). With the laser scanning microscope and the dark field luminescence microscope we can observe living organelles of submicrometer size. With there atomic-scale images we felt the coming of the age of nanotechnology age, well illustrated by the STM image of single-stranded DNA on the cover of *Nature* (Dunlap and Bustamante 1989). In

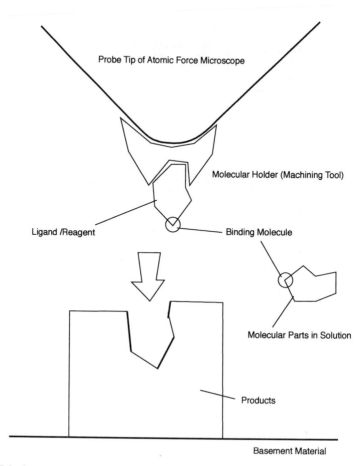

Fig. 7.9   In water, reactive molecules are arranged with atomic precision by an atomic force microscope.

1990, a DNA double helix was observed on an atomic scale with an atomic force microscope (AFM) (Driscoll *et al.* 1990).

Many researchers immediately thought that it might be possible to use SPMs as the manipulator for handling molecules and atoms (Fig. 7.9). A technique for stereoscopic handling of microstructures in a electron scanning microscope has already been developed (Hatamura and Morishita 1990). The famous 'IBM' nameplate which was written with 32 xenon atoms on an Ni plate under super high vacuum using an AFM probe was a good example of atomic manipulation

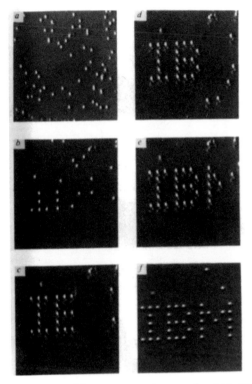

Fig. 7.10 A name plate formed by a patterned array of xenon atoms on a Ni (110) surface. The image was taken by STM and formed by an AFM under super high vacuum. Each letter is 50 Å from top to bottom. (From Eigler and Schweizer 1990.)

(Fig. 7.10). In Japan, Hitachi, JEOL, NTT, ERATO JRDEC, etc., have produced many kinds of atomic scale characters and images using SPMs or scanning electron microscopes. In biology, an intercellular manipulator has been proposed and developed by Higuchi *et al.* (1990).

Table 9.1 Biomimetic approaches in inorganic materials chemistry (adapted from Mann (1993))

Approach	Strategy	Product	Systems	Materials
Nanoscale synthesis	Host–guest	Clusters	Reverse micelle	CdS
			Microemulsions	Pt, Co metal borides
				Fe_3O_4, $CaCO_3$
		Nanoparticles	Vesicles	Pt, Ag
				CdS, ZnS
				Ag_2O, FeOOH, Fe_3O_4, Al_2O_3
				Ca phosphates
				MnOOH, UO_3, FeS, Fe_3O_4
			Ferritin	CdS
			LB films	γ-Fe_2O_3
	Ligand capping		Polystyrene resin	CdS
			(γ-EC)nG peptides	
Crystal engineering	Oriented nucleation	Single crystals	Monolayers	NaCl, $CaCo_3$, $BaSO_4$, PbS
			Polyasp/polystyrene	$CaCO_3$
	Templating	Shaped composites	Tubules	Cu, Ni, Al_2O_3, Fe oxides
			Bacterial fibres	Fe_2O_3, $CaCO_3$, CuCl
			Bacterial rhapidosomes	Pd
			S-layer proteins	Ta/W
	Directed growth	Textured crystals	Sea-urchin proteins	$CaCO_3$
			Polyanionic peptides	$CaCO_3$
Microstructural fabrication	Extended frameworks	Mineral–polymer composites	Polystyrene-butadiene	$CaCO_3$, CdS, Ca phosphate
			Polyvinylchloride	TiO_2
			Polyacrylate films	Fe oxides, $BaTiO_3$
			Polysiloxanes	$CaCO_3$, $CaSiO_3$
			Polyethylene oxide	CdS
			Collagen	Ca phosphate
	Assembly	Organized materials	Monolayers/Au	CdS
			Monolayers	Fe_2O_3
			Cast bilayers, films	Fe_3O_4
			Hydroxyethyl cellulose	$CaCO_3$
			Polyacrylate sols	$BaSO_4$
			SiO_2/OH- gel	$CaCO_3$

The allometric relations of efficiency or cost of the minimum speed are ruled by dynamics concerning the animal's size and characteristics its environment. The costs of locomotion and body weights are inversely proportion when plotted in logarithmic coordinates (Vogel 1988). For free movement of animals, if the rate of energy consumption per unit basal metabolic rate affects the allometric relation, the rates might not change with the body weight. Actually, the rate of energy consumption of animals in water is almost the same whatever the body weight. Cold-blooded animals in water expend only the basal metabolic rate for locomotion and warm-blooded animals in water expend twice the basal metabolic rate. But for birds and warm-blooded animals on land the rates markedly increase nonlinearly as the body weight increases. These data indicate that the 'engine' structure is different in animals that run fast (Peters 1983); in the same way the engine horsepower of a car is not proportional to the car's weight but to its speed.

8.2 Engines of biological systems

From allometric analysis, we can imagine that there might be some common energy conversion elements in living things. We know that the mechanisms of motive force (engines) of animals are of only three kinds: muscles, cilia, and flagella in eukaryotic cells. There are also bacterial flagella. These mechanisms have their functional limits according to the size of the organism. Muscle exists in animals longer than 0.1 mm, cilia in animals with body length 20 μm to 20 mm, and flagella in those with body length 1 μm to 50 μm. The velocities of the engines and the size of engine systems might be proportional it expressed logarithmically, as shown in Fig. 7.2 (Fujimasa 1994).

Biological engines are composed of motor proteins and cytoskeletons. The basic mechanisms of motive force occur between the cytoskeleton and the motor proteins. The size of a motor protein is almost 100 nm and one power stroke produces a force of about 1 pN. This force has recently become measurable, but its precise energy conversion mechanism is still unknown.

9
Artificial life

9.1 When did living organisms become machines?

On Earth, life appeared about 3.5 billion years ago. Until two billion years ago, living things remained unicellular systems. Prokaryotic cells were filled with water soluble materials and the system looked like a chemical machine. Long after the prokaryotic age, some prokaryotes became parasites for other cells and organelles of eukariotic cells. About 1.5 billion years ago, the main motor protein and cytoskeletons were designed and organelles, motor proteins, and cytoskeletons formed cell structure. At the end of the age, the cell had become a mechanical machine system.

About five hundred million years ago, in the pre-*Cambrian*, the first multicellular animals appeared in the sea. This is usually called the pre-*Cambrian* explosion of life. Clear evidence for the existence of many unusual animals, almost none of which can be classified in terms of today's animals, were obtained from the Burgess shale in the Canadian Rocky Mountains: *Opapinia* with five eyes, the flat shoehorn-like animal *Odontogrifis*, *Anomarocaris* of large carnivorous arthropods, *Hallucigenia* with seven mouths and unbelievable outer shape. This seems to be the first stage in the design of multicellular animals and freedom of design allowed the existence of many such types of animals.

With man-made machines, the same phenomena appears as the first prototype model of newly developed machines. Freedom of design concepts are almost infinite at that stage. Almost all the essential parts for biological machines already existed in the pre-*Cambrian* period. Using such parts, life designed its own structure with complete freedom. At this period, animals became mechanical machines. But the dynamic principle might have been completed one billion years earlier. Around a billion years ago, the DNA (fabrication charts) of many mechanical parts of eukaryotic cells had already been 'designed'. Many motor proteins such as kinesin, dynein, and myosin had appeared in eukaryotic cells. Cytoskeleton structures such as F-actin, microtubules, and intermediate fibers already existed. Thus, at around one billion years ago the cell was already constructed as a machine. These cells used ATPs as energy sources, producing energy mainly by oxidizing glycolysis on the mitochondrial membrane. The 'machine' to produce ATP, called ATP-ase, already existed and its structure was similar to a turbine converting energy from proton (H^+) ion flow pumping it up to ATP. Many kinds of fundamental mechanochemical parts were developed

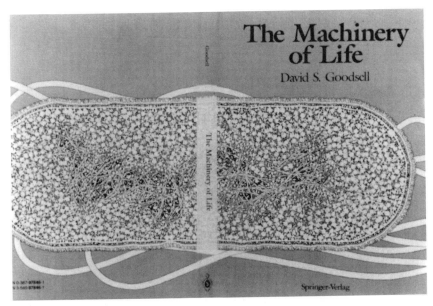

Fig. 9.1   This book by Goodsell (1994) described a bacterium as a molecular machine. Some eukaryotic cell structures are also illustrated.

between two billion and one billion years ago (Fig. 9.1). The mechanism of energy conversion from chemical potential to mechanical work would have been fundamentally fixed in that era. However, we cannot yet clarify the mechanism.

9.2   Artificially structured materials

Recently, a breakthrough has been made in mesoscopic machining in the field of structural technology. Many materials scientists have begun to survey the structure of hard materials in living things. Examples of such materials are intercellular magnetic (Fe_2O_3) crystals in magnetotactic bacteria, intracellular $CaCO_3$ crystals formed within the *Coccolithophoride syracosphaera* or in fish bones, and SiO_2 rods in some eukaryotic cells (Fig. 9.2). Then, some materials scientists fabricated a ferrofluid containing nanoscale magnetic crystals formed within the cavity of individual ferritin molecules, an array of oriented calcite crystals, and an organized aggregate of $BaSO_4$ crytals formed from an aqueous supersaturated solution. These products have a similar structure to inorganic materials in living bodies (Table 9.1). This research is called 'molecular tectonics' by Douglas and Mann (1993).

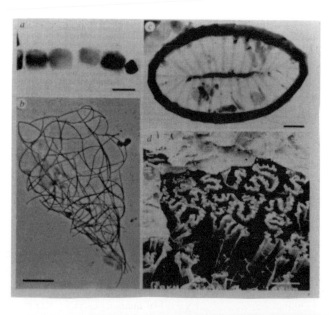

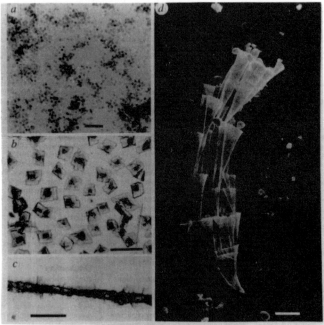

Fig. 9.2   Molecular tectonics of in organic crystals. 1 (upper) shows crystals produced in living things: (a) Fe_2O_3; (b) $CaCO_3$; (c) SiO_2; and (d) $CaCO_3$. 2 (lower) shows crystals produced by autoassembly in aqueous solution: (a) Fe_2O_3; (b) $CaCO_3$; (c) Fe_2O_3; (d) $CaCO_3$.

8
The cell as a basic element of life

Biological systems are good examples of micromachines. In this chapter, I shall describe the mechanics of living organisms in terms of machines.

8.1 The mechanics of life

A living thing is made up of cells. The cell is a basic element of life. If we view life as a machine constructed with the same structural elements, the energetics of life should obey a simple rule proportional to body size. Actually, the logarithmic value of basal metabolic rate is proportional to that of body weight from large animals to unicellular animals. The method of analysis used for size-related items is called allometry. The allometric relation between basal metabolic rate (E_s) and body weight (W) is shown in Table 8.1. The relationship indicates that animals are machines constructed with common metabolic parts from unicellular animals to large warm-blooded animals.

The system of production of motive force in animals might also obey the allometric principle. As feeding rates (I) of animals, which depend upon the basal metabolic rates, also follow the allometric principle, their speed of movement, which is related to their ability to acquire food, obeys the allometric relation. We can find one allometric formula for insects and mammals running on land and another for fish and flying birds. It is clear that animal motive force is produced by a system consisting of common actuating elements, and the allometric formulae only differ due to their differing environments.

Table 8.1 Allometry of the basal metabolic rate (E_s) to body weight (W) (Motokawa 1993)

Animals	Basal metabolic rates $E_s(W)$	Feeding rate $I(W)$
Warm-blooded animals	$4.1W^{0.751}$	$10.7W^{0.703}$
Cold-blooded animals	$*0.14W^{0.751}$	$0.78W^{0.82}$
Unicellular animals	$*0.018W^{0.751}$	

$*E_s$ of cold-blooded and unicellular animals were measured in 20 °C.

9.3 Artificial life

Humans use their brains to produce new information and knowledge. Today, artificial products including software and hardware reproduce themselves. This self-replication mechanism has many similarities to living systems. In a computer we can produce self-replication mechanisms such as computer viruses and many other autonomous things similar to living things. Today, we called them 'artificial lives'. The concept of 'artificial life' firstly appeared among hackers and it means living things in computer system, but today, it means a man-made thing or system that has similar functions to living things. In computer science many software techniques have learnt from the characteristics of living things have been developed since the 1960s and have become key techniques in artificial intelligence. 'Adaptive system theories' have simulated the ability to adapt of living things, and 'evolutionary computation' has simulated algorithms of reproduction and evolution in living things. Following these trends the mechanism of gene replication has been analysed and made into a software tool called a 'genetic algorithm'. Today the software is only used to analyse complex systems, but in future, it might include functions of self-assembly machine design.

10
Mesoscopic mechanical engineering

10.1 Mechanics of mesoscopic machines.

Behind micromachine technology we find nanotechnology, which would support true micromechanical systems. If we want to develop micrometer machines, they should integrate several elements of nanometer size. If we were to produce a soft actuator fitting any shape like a muscle, the size of its basic element might be in nanometers. This area has recently been called the mesoscopic domain, and such engineering is called mesoscopic mechanical engineering.

As an example, if we develop an implantable artificial heart we should try to find some integrated actuators similar to heart muscles, consisting of a large number of sarcomeres. A sarcomere is a large-scale integrated structure of actin–myosin complexes. On such a scale, the dynamics for converting energy obey quantum physics including thermal molecular vibration, called Brownian motion. Moreover, many conventionally structured micromachines cannot move using their own actuators, because frictional or viscous forces exceed the actuator driving force. However, living things can move in such conditions using a combination of motor proteins and cytoskeletons. We must construct a new principle for energy conversion in the mesoscopic domain and find a technology which can be used to fabricate mesoscopic machines.

10.2 Dynamic design rules of biological actuators

Biological actuators, such as the myosin–actin complex and the dynein–tubulin system, use high-energy phosphate as their energy source. ATP (adenosine triphosphate) is the source of the phosphate. When 1 mole ATP is hydrolysed, around 10 kcal of energy is released. Therefore, when an ATP molecule is converted to ADP (adenosine diphosphate), the energy of a high-energy phosphate molecule is

$$E_{\text{ATP}} = 10 \text{ kcal}/N_{\text{A}} = 6.96 \times 10^{-20} \text{ J}$$

where N_{A} is the Avogadro number. If the total energy of a high-energy phosphate is used in moving a G-actin molecule, which is a component of an actin bundle, the initial velocity is calculated as follows:

$$E_{ATP} = \frac{1}{2} m_{G\text{-actin}} v_{G\text{-actin}}^2$$

where

$$m_{G\text{-actin}} = 4.2 \times 10^5 \text{ Da} = 6.97 \times 10^{-22} \text{ kg}$$

$$(\text{atomic mass unit} = 1.66 \times 10^{-27} \text{ kg})$$

and

$$v_{G\text{-actin}} = 44.8 \text{ m s}^{-1}.$$

Meanwhile, the average kinetic energy for independent motion (one degree of freedom) is considered to be the energy of Brownian motion (E_B) and is calculated as

$$E_B = \frac{1}{2} k_b T,$$

$$k_b = 1.38 \times 10^{-23} \text{ J K}^{-1} \text{ (Boltzmann's constant)},$$

$$E_B = 2.14 \times 10^{-21} \text{ J (in three degrees of freedom, } 6.42 \times 10^{-21} \text{ J).}$$

If we compare the energy of thermal motion of a Brownian particle and the energy of hydrolysis of ATP, the ratio E_{ATP}/E_B is around 11. This result means that the energy obtained from ATP hydrolysis is at most 10 times greater than the thermal noise energy. Thus it is clear that the actuator in a biological system uses energy at the level of thermal noise. We can obtain the mean square velocity of a Brownian particle at 37 °C, the mass of which is 10^{-15} kg (m_B: the mass of 1 μm^3 of water), using the following formula:

$$E_B = \frac{1}{2} m_B v_x^2$$

which gives

$$v_x = 2.10 \times 10^{-3} \text{ m s}^{-1} \text{ (independent direction)}$$

or

$$v = 3.58 \times 10^{-3} \text{ m s}^{-1} \text{ (three degrees of freedom).}$$

How far can a microparticle move without collision? As the mean free time of Brownian particles (τ_B) is estimated as 10^{-8}–10^{-9}, the mean free path becomes 10^{-11}–10^{-12} m (10^{-1} picometer (pm)). On the other hand, the mean free time of a G-actin molecule ($\tau_{G\text{-actin}}$) can be calculated from $\tau = m/6\pi R\eta$ (see eqn (10.6) and preceding discussion) and is 8.55×10^{-13} s ($m = 6.97 \times 10^{-23}$ kg; $R = 4.0 \times 10^{-9}$ m; $\eta = 1.08 \times 10^{-3}$ kg m^{-1} s^{-1}). Therefore, a G-actin molecule can move only 38 pm with its initial speed in one direction. The reason why the mean free path of a G-actin molecule is so short is due to the disturbance of Brownian motion and the viscous drag of water molecules.

One can therefore understand that biological actuators are influenced by thermal molecular motion. Yanagida observed *in vitro* that a 40 nm long actin filament, that is the shortest unit of an actin fliament for gliding and using hydrolysis of one ATP molecule for gliding, moved at 6 μm s^{-1} on a myosin-coated plate. In his experiment he used a cover glass coated by myosin monomers with a close-packed structure and observed the gliding speed of a single actin filament on the plate under load-free conditions. He estimated that the 40 nm long actin filament moved 60 nm with an ATP energy input. This result means that an actin filament can travel 2000 times further than estimated with the energy of a single ATP hydrolysis. In theory we assumed the mean free time of the 40 nm actin filament to be only 10^{-11} s, but the real mean free time is 0.01 s. From this result we must conclude that the actin–myosin complex is somehow protected from molecular movement of water due to thermal agitation.

Large-scale integration is a solution. How many pairs of actin–myosin complexes do we integrate in order to increase the mean free path (X_A) to the actual actin gliding length using the energy of a single ATP hydrolysis? Let us try to calculate the mass (M) and the number of integrations. In order to calculate the virtual mass we use the following data: the mean free path of an actin–myosin complex (X_A) is 60 $\times$ 10^{-9} m; density of the actin–myosin complex (P) is around 1.0 $\times$ 10^3 kg m^{-3}; the velocity of actin movement is V; the energy obtained from ATP hydrolysis (E_{ATP}) is 6.96 $\times$ 10^{-20} J; the viscosity of water at 37 °C (η) is 10^{-3} kg m^{-1} s^{-1}; and the virtual radius of an actin–myosin complex is R. Using the following three equations:

$$X_A = V_\tau = MV/6\pi R\eta,$$
$$M = 4\pi R^3 \rho/3,$$
$$E_{ATP} = MV^2/2,$$

we find

$$M = \left[(6\pi^{2/3}\eta X_A)(3/4d)^{1/3}/(2E_{ATP})^{1/2}\right]^6 = 4.34 \times 10^{-5} \text{ kg}.$$

When we consider that the 40 nm actin filament consists of 15 to 16 molecules of G-actin and the length is almost the same as that of the half-pitch of an α double helix of the actin filament (36 nm), the weight is at most 10^{-21} kg. There is a large discrepancy between the theoretical mass of an actin filament moving 60 nm and the real weight of a 40 nm actin filament. If we calculate the weight of the actin–myosin complex, which is part of a sarcomere, it is only 2.87 $\times$ 10^{-18} kg and it cannot reach the weight of a Brownian particle (10^{-15} kg). In order to move 60 nm as a mean free path using the energy of hydrohysis of ATP, the muscle should be made of 10^{13}–10^{14} integrated actin–myosin complexes. Otherwise, the sarcomere should have some structure for protecting it from external thermal agitation, or have some electrostatic gradient between the actin and myosin filaments. Anyway, integration is the essential design rule for biological actuator systems.

10.3 Physics of the micromechanism: the Langevin equation

The mathematical model indicates some design principles for mesoscopic machines and makes it possible to compare the characteristics of biological actuator elements and artificial mesoscopic actuators.

When we describe the dynamics of micromechanisms, the microstructure obeys the following rules in the order of size:

Macroscopic micromachine (≥ 1 μm): classical Newtonian physics,
Mesoscopic micromachine (10 nm $-< 1$ μm): turning point from classical
   Newtonian physics to quantum physics,
Atomic and molecular machines (< 10 nm): quantum physics.

I want to summarize the equations of motion from macroscopic to quantum scale (Table 10.1). When we consider particles in water, each equation is as follows. Thermal molecular motion (thermal agitation) around an actuator gradually influences its movement as the actuator becomes smaller. However, we can apply Newton's second law of motion for analysing its dynamics down to a size of 1 μm. It is possible to apply scaling analysis. Newton's equation of motion is written as follows:

$$m \frac{dv}{dt} = -\beta v + K(t) \tag{10.1}$$

where βv is the viscous drag force and $K(t)$ is the external force. The viscous dragging force is the mean viscous frictional force that occurs due to collision with water molecules.

The smaller the size of machine becomes, the larger the viscous dragging force. Moreover, we cannot neglect the Brownian motion of water molecules. When the size of a micromachine is less than 1 μm, the size becomes comparable to that of Brownian particles. For example, the Brownian motion of water influences the movement of actin–myosin complexes, the most important actuator element of biological systems. Moreover, some physiologists suspect that the actin–myosin complex converts the thermal agitation energy into motion (Vale 1993). Therefore, we must take the effect of the thermal agitation into consideration when analysing mesoscopic actuators. The equation that adds the stochastic force of thermal agitation to Newton's equation of motion is called the Langevin equation:

$$m \frac{dv}{dt} = -\beta v + F(t) + K(t). \tag{10.2}$$

In this formula $-\beta v$ is the viscous drag force, i.e. the average frictional force caused by molecular thermal motion. Therefore, β relates to the frictional coefficient μ as:

Table 10.1   Relations between scale dimensions and equations of dynamics

Scale	Dominant phenomena	Fundamental equations
10^{-3} m Mesoscopic domain	Macroscopic dynamics Microscopic dynamics	Newton's equations of motion
10^{-6} m	Brownian motion field	Introduction of fluctuation Langevin equation
	Two inertial forces caused by large acceleration	Introduction of expanded Langevin equation
	Application of molecular dynamics Supermolecules Influences of water particle movement	Newly expanded Langevin equation
	Shift to quantum physics	Expanded Langevin–Schrödinger equation
10^{-9} m	Quantum physics of many-body problem	Quantum statistical dynamics
10^{-12} m	Interaction of elementary particles	Quantum theory of field

$$\beta = 1/\mu = m/\tau \qquad (10.3)$$

where m is the mass of a Brownian particle and τ is the mean scattering time. $F(t)$ is called the stochastic force that is generated by collision with Brownian particles, i.e. water molecules, and fluctuates around the viscous drag force. We can express the sum of viscous drag and stochastic forces as the frictional force in the Brownian domain. $K(t)$ represents external forces such as gravitational force, electromagnetic force, etc.

In order to analyze the dynamics of a micromechanism in the Brownian domain, which means the size of the mechanism is around 1 μm, we use the following assumptions:

1. The ensemble mean of the stochastic force is equal to zero: $F(t) = 0$.
2. Successive collisions are dynamically correlated with the stochastic force; when the interval between collisions become very short the correlation is

written as follows:

$$F(t_1)F(t_2) \equiv \phi(t_1 - t_2) = C\delta(t_1 - t_2) \qquad (10.4)$$

and $\qquad C = 2k_B T \beta$

where k_B is Boltzmann's constant. In this formula, $F(t)$ has a high amplitude at around $t = 0$. Therefore, we can approximate the formula to a delta function. This process is also a Markov chain process, in which only the final collision influences the dynamics of the present collision.

3. The shape of object is assumed to be a ball with radius R. If that assumption is not possible, we take R as the fluid dynamical radius. Then we can apply Stokes' law to the object:

$$\beta = 6\pi R\eta \quad \text{(Stokes' law)}$$
$$f = -6\pi R\eta v = \beta v \qquad (10.5)$$

where f is the force due to viscosity and η is the coefficient of viscosity. From equations (10.3) and (10.5) we have

$$\tau = m/6\pi R\eta. \qquad (10.6)$$

We can use the Langevin equation as a basic dynamic equation for analysing the motion of a simple object of around 1 µm in length.

For example, the motion of a saw device was calculated by Matuura (1993). A slider shaped like a saw and a stator having saw teeth in the form of a 1 µm triangle are set in water and each saw is made to vibrate forcibly so that they collide with each other. According to the strength of impact forces applied on each edge of the saw-teeth triangles that have asymmetric base angles, the sliders of the saw device move in one direction. The maximum speed of motion and the most effective angular speed for its actuating vibration were calculated using the Langevin equation (Fig. 10.1). Matuura tried to develop an artificial sarcomere element actuated by a vibrational force field, and to expand the analysis into the mesoscopic domain (see Fig. 7.6).

10.4 Dynamics in the mesoscopic domain: the expanded Langevin equation

The Langevin equation does not consider the duration of a collision between target particles, that we call mobile objects, and Brownian particles, but this duration influences the speed of actuator motion. If the mobile object is smaller than 1 µm and is hard, it markedly accelerates after collision, and as a natural consequence the inertial resistance cannot be ignored. If we compare the Newtonian inertial

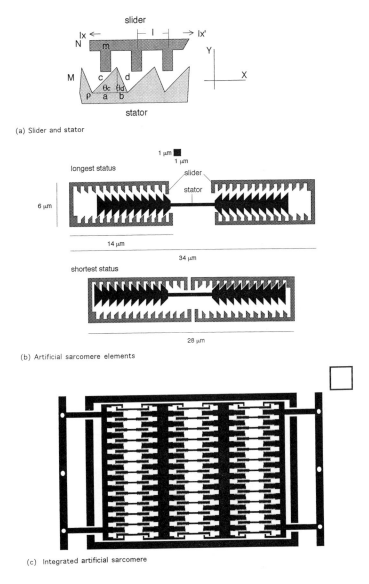

(a) Slider and stator

(b) Artificial sarcomere elements

(c) Integrated artificial sarcomere

Fig 10.1 Design of artificial sarcomere unit. A slider shaped like a saw and a stator having saw teeth is made to vibrate forcibly so that they collide with each other. The slider moves in one direction according to the strength of impact forces. A fabricated example is shown in Fig. 7.6. (a) Principle of collision between a stator and a slider. (b) Design of a final model. (c) An integrated sarcomere model.

resistance $(Q(t) = \pi R^2 \rho v^2/4)$ (called inertial resistance 2) with the viscous resistance $(\beta v = 6\pi R \eta v)$, the inertial resistance becomes 10^6 times larger and the viscous resistance 10^3 times larger when the collision time becomes 10^3 times smaller. As the duration of the collision is thought to be prolonged in soft materials, cellular mechanical components, such as motor proteins and cytoskeletons, have longer collision times. In such conditions, we must take the influence of inertial resistance 2 into consideration for the equation of motion. Meanwhile, a rapid increase in acceleration of the mobile object gives rise to many dynamical influences upon the surrounding media. This influence is expressed by increasing virtual mass of the mobile object. The inertial resistance is expressed as $\frac{2}{3}\pi R^3 \rho(dv/dt)$. This inertial resistance is called inertial resistance 1. This is effective at very early stages of the collision and we can usually observe the effects of inertial resistance 2. Here we call the equation that includes those two inertial resistances the expanded Langevin equation (ELE):

$$\left(m + \frac{2}{3}\pi\rho R^3\right)\left(\frac{dv}{dt}\right) = -\beta v + F(t) + Q(t) + K(t)$$

where $\frac{2}{3}\pi\rho R^3\left(\frac{dv}{dt}\right)$ is inertial resistance 1

and $Q(t) = \pi R^2 \rho v^2/4$ is inertial resistance 2.

Using the ELE, we can analyse the dynamics of actuators 1 μm long.

11
Mesoscopic mechanical engineering: the next engineering era

11.1 Conclusion

In engineering we have developed many machines from experience and from new ideas. There have been two types of machine: one has been produced by trial and error, the other from theoretical design dependent upon dynamics. However, we now try to design new machines with a new philosophy. In this book I have written about a new method of designing machines. The principle depends upon the physics of the submicrometer domain, the kingdom of biological machines. Let us explore the kingdom. We need technological advances to enter the kingdom, and find ourselves in the same situation as Leonardo da Vinci in the Renaissance—we can speculate upon the designs of our machines but cannot yet build them.

Glossary

Amino acid: Amino acid molecules are the building blocks of proteins. They contain both an amine and a carboxylic acid group; in the 20 genetically encoded amino acids in biology, both groups are bound to the same carbon. Amino acids are connected to each other by amide bonds to form peptides and proteins.

Anisotropic etching: An etching technique in which grooves are sculpted in a silicon substrate with a specific crystal orientation. The basic etchant is KOH.

Assembler: A programmable nanomachine which uses an assembler to perform a wide range of mechanosynthetic operations.

Atomic force microscpe (AFM): A device in which the deflection of a sharp stylus mounted on a soft spring is monitored as the stylus is moved across a surface. If the deflection is kept constant by moving the surface up and down by measured increments, the result (under favorable conditions) is an atomic-resolution topographic map of the surface. Also termed a scanning force microscope.

Autoassembly: A fabrication process typically seen in living cells. However, it is commonly called *self-assembly* in biotechnology. A machine can be programmed to build any structure or device from simpler elements. It is analogous to computer-aided manufacturing.

Brownian assembly: A method of assembly using Brownian motion. Brownian motion in a fluid brings molecules together in various positions and orientations. If molecules have suitable complementary surfaces they can bind, assembling to form a specific structure.

Bulk micromachining: A fabrication technique for making microstructures, it is used in the silicon batch process based on sculpting the silicon substrate with a chemical etchant and forming silicon microstructures in orientation-dependent etching solutions.

Electrodischarge machining: A micromachining technique using electrical discharge. In the microdomain this technique has the great advantages of forming only low voltages and enabling precision manufacturing.

Electrostatic actuator: An actuator driven by electrostatic motive force. In the microdomain its power has a great advantage as a microactuator, because the power scales proportional to the square.

Flagellar motor: An ultra-small motor on the surface of a bacterial cell.

Internal energy: The sum of the kinetic and potential energies of the particles that make up a system.

Isotropic etching: An etching technique involving sculpting on a silicon substrate without a specific crystal orientation. The technique is usually used for sacrificial layer etching. The basic etchant is hydrofluoric acid (HF).

Lateral aspect ratio: In surface micromachining, the ratio of the length of a structure in the plane of the wafer to its width in that plane.

LIGA process: LIGA stands for the German equivalent of 'lithography, electroforming, and moulding'. The source energy for the LIGA process is a short-wavelength X-rays produced by synchrotron orbital radiation.

Microactuator: A fundamental component of a microdynamic system. For example, the micromotor, the linear microactuator, and the microvibrator have been developed as microelectromechanical systems.

Microelectromechanical system (MEMS): A miniature device or an array of devices combining electrical and mechanical components fabricated using IC batch-processing techniques.

Micromachine: A machine made from components (mechanical, electronic, chemical, thermal, or otherwise) on a micrometer scale, also known as 'micromechanical devices', 'microdynamical systems', and 'micromechanisms'.

Molecular machine: A mechanical device that performs a useful function using components of nanometric scale and defined molecular structure which includes both artificial devices and biological systems.

Molecular manufacturing: The production of complex structures using non-biological mechanosynthesis.

Molecular nanotechnology: A technology based on the ability to build structures to complex atomic specifications by means of mechanosynthesis. The term was defined by Drexler.

Nanomachine: A machine made from components (mechanical, electronic, chemical, thermal, or otherwise) on a nanometer scale.

Nanotechnology: In recent general usage, any technology related to features of nanometer scale: thin films, fine particles, chemical synthesis, advanced microlithography, etc.

Organelles: Parts making up a cell structure. For example, the sarcomere is equivalent to the linear actuator, the microtubule to the belt-conveyor in the cell, the mitochondrion to an energy plant, and the ribosome to the tape reader of RNA.

Piezoelectric actuator: An actuator applying the piezoelectric effect. The actuators are applied to an instrument with nanoprecise actuation, such as the actuator of a scanning tunneling microscope and micromanipulated injector.

Piezoelectric effect: A phenomenon in which certain materials become electrically polarized in response to applied strain or become strained in response to applied voltage.

Piezoresistor: A resistor that changes its resistance in response to applied strain.

Proof mass: The reference mass inside an accelerometer, from whose movement the acceleration measurement is derived.

Protein technology: Technology based on the manipulation of amino acids and other useful biological molecules to build proteins using biological cells and modified genes. Protein technology is the fundamental fabrication technology for biotechnology.

Released mechanical layer: A structure that is separated from the substrate by the thickness of the sacrificial layer.

Sacrificial layer: The most fundamental technique of the silicon batch process that produces a free-moving structure. The layer films are grown on a silicon substrate by low-pressure chemical vapor deposition and etched by HF. As the result, the deposited layer on the sacrificial layer remains as a movable structure.

Sarcomere: A basic actuator element of striated muscle. It is composed of some protein filaments such as actins (structural backbones: cell skeleton), myosins (motor elements: a kind of motor protein), and z-discs (end plate).

Scanning tunneling microscope (STM): A device in which a sharp conducting tip is moved across a conducting surface close enough to permit a substantial tunneling current. The STM has been used to manipulate atoms and molecules on surfaces.

Self-assembling: See *autoassembly*.

Sensor: A device providing useful output to a measuring instrument, but whose value may depend on more than one variable.

Shape memory alloy (SMA): An alloy that can crystallographically return to its original shape memorized at high temperature.

Silicon batch process: A fabrication process that integrates many electronic and mechanical parts on one piece of silicon wafer using crystal growth and lithography technologies. The smaller sizes of the parts are typically smaller than 1 µm.

Silicon fusion bonding: A process for bonding two silicon wafers at the atomic level without applying glue or an electric field.

Smart sensor: A sensor device with built-in intelligence performed by circuits on a chip.

Structural layer: A layer of material deposited in the surface micromachining process that will become a structural member of a microstructure, in contrast to a sacrificial layer.

Surface micromachining: A process for depositing and etching multiple layers of sacrificial and structural thin films to build complex microstructures.

Thermal energy: The internal energy present in a system as a result of thermally equilibrated vibration modes and other motions. The mean thermal energy is kT.

Thermal fluctuation: The thermal energy of a system has a mean value determined by the temperature and the structure of the system. Stastistical deviations about that mean are thermal fluctuations. These are of great importance in determining the dynamics of nanomachines.

Transducer: A device that converts energy from one form to another, calibrated to minimize errors in the conversion process.

Transmitter: A transducer with a standard output format suitable for a long-distance transmission.

Tribology: The study of friction, wear, and lubrication in surfaces sliding against each other, as in bearings and gears.

Van der Waals force: A kind of intermolecular attractive force not resulting from ionic charges.

Vertical aspect ratio: The ratio of the height of a structure perpendicular to a wafer's surface to its thickness in the wafer's plane.

Young's modulus: A modulus relating tensile stress to strain in a rod. The relevant measure of strain is the elongation divided by the initial length.

References

(1987). Stereolithography, unveiled. *Newsweek*, p. 5.

(1990). *1990 International Micromachines Symposium (Simon Frazer University, Burnaby, British Columbia)*. SPARK, Science Council of British Columbia.

(1992). *Proceedings of IEEE Microelectromechanical systems: an Investigation of Microstructures, Sensors, Actuators, Machines, and Robotic Systems, (Travemünde, Germany)*. IEEE, New York.

(1993). *Proceedings of IEEE Microelectromechanical Systems: an Investigation of Microstructures, Sensors, Actuators, Machines, and Robotic Systems (Fort Lauderdale, FL)*. IEEE New York.

(1994). *Proceedings of IEEE Microelectromechanical Systems: an Investigation of Microstructures, Sensors, Actuators, Machines, and Robotic Systems (Oiso, Japan)*. IEEE, New York.

Abela, G.S *et al.* (1985). Hot-tip: another method of laser vascular recanalization. *Lasers in Surgery and Medicine*, **5**, 327–35.

Ahn, C.H., Kim, Y.J., and Allen, M.G. (1993). A planer variable reluctance magnetic micromotor with fully integrated stator and wrapped coils. In *IEEE Microelectromechanical Systems (Fort Lauderdale, FL)* pp. 1–6. IEEE, New York.

Akin, T., Ziaie, B., and Najafi, K. (1990). Telemetry powering and control of hermetically sealed integrated sensors and actuators. In *Technical Digest of the IEEE Workshop on Solid-State Sensors and Actuators (Hilton Head, SC)*. IEEE.

Akin, T., Najafi, K., Smoke, R.H., and Bradley, R.M. (1994). A micromachined silicon sieve electrode for nerve regeneration applications. *IEEE Transactions on Biomedical Engineering*, **41**, 305–13.

Amato, I. (1992). A new kind of organic gardening. *Science*, **258**, 1084.

Ando, J., Nomura, H., and Kamiya, A. (1987). The effect of fluid shear stress on the migration and proliferation. *Microvascular Research*, **33**, 62–70.

Ashkin, A. (1980). Applications of laser radiation pressure. *Science*, **210**, 1081–90.

Barnard, S.M. and Walt, D.R. (1991). A fibre-optic chemical sensor with discrete sensing sites. *Nature*, **353**, 338–40.

Bassous, E. (1978). Fabrication of novel three-dimensional microstructures by the anisotropic etching of (100) and (110) silicon. *IEEE Transactions on Electron Devices*, **25**, 1178.

Bean, K.E. (1978). Anisotropic etching of silicon. *IEEE Transactions on Electron Devices*, **25**, 1185.

Becker, E.W., Ehrfeld, W., Münchmeyer, D., and Betz, H., *et al.* (1982). Production of separation-nozzle systems for uranium enrichment by a combination of x-ray lithography and galvanoplastics. *Naturwissenschaften*, **69**, 520–3.

Behi, F., Meheregany, M., and Gabriel, K.J. (1990). A microfabricate three-degree-freedom parallel mechanism. In *IEEE Microelectromechanical Systems (Napa Valley, CA)*. pp. 159–65. IEEE, New York.

Benecke, W. (1989*a*). Grundstrukturen und Elemente der Mikromechanik. In *Mikromechanik* (ed. A. Heuberger), pp. 343–54. Springer, Berlin.

Benecke, W. (1989*b*). Anwendungen mikromechanischer Bauelemente und Komponenten. In *Mikromechanik* (ed. A. Heuberger), pp. 355–418. Springer, Berlin.

Benecke, W. (1989*c*). Bauelemente für konstruktive Probleme in verschiedenen Bereichen der Technik. In *Mikromechanik* (ed. A. Heuberger), pp. 419–31. Springer, Berlin.

Benecke, W. (1989*d*). Integration von Mikromechanil und Mikroelektronik auf einem Siliziumchip. In *Mikromechanik* (ed. A. Heuberger), pp. 329–42. Springer, Berlin.

Bhushan, B., Israelachvill, J.N., and Landman, U. (1995). Nanotribology: friction, wear and lubrication at the atomic scale. *Nature*, **374**, 607–16.

Binning, G. and Rohrer, H. (1985). The scanning tunneling microscope. *Scientific American*, **253**, (August), 50–6.

Biresaw, G. (ed.) (1990). *Tribology and the liquid-crystalline state*. ACS Symposium Series vol. 441.

Block, S.M. (1992). Making light work with optical tweezers. *Nature*, **360**, 493–5.

Bobbio, S.M., Kellam, M.D., Dudley, B.W., and Goodwin-Johansson, S., *et al.* (1993). Integrated force arrays. In *IEEE Microelectromechanical Systems* (*Fort Lauderdale, FL*), pp. 149–154. IEEE New York.

Branebjerg, J., Eijkel, C.J.M., Gardeniers, J.G.E., and van der Pol F.C.M., (1991). Dopant selective anodic etching of silicon. In *IEEE Microelectromechanical Systems* (*Nara, Japan*) pp. 221–26. IEEE, New York.

Brennen, R.A., Pisano, A.P., and Tang, W.C. (1990). Multiple mode micromechanical resonators. *Proc. IEEE Microelectromechanical Systems* (*Napa Valley, CA*), pp. 9–14. IEEE, New York.

Brooks, R.A. (1987). Micro-brains for micro-brawn; autonomous microbots. In *IEEE Micro Robots and Teleoperators Workshop*, (*Hyannis, MA*) p. 27. IEEE, New York.

Bryzek, J., Petersen, K., and McCulley, W. (1994). Micromachines on the march. *IEEE Spectrum Magl.* **494**, 20–31.

Buehler, W.J., Gilfrich, J.W., and Weiley, K.C. (1963). *Journal of Applied Physics*, **34**, 1467.

Cabuz, C., Shoji, S., Fukatsu, K., and Cabuz, E., *et al.* (1993). Highly sensitive resonant infrared sensor. In *Transducers '93* (*Yokohama, Japan*), pp. 694–7.

Chaw, H.L. and Wise, K.D. (1988). *IEEE Transactions on Electron Devices*, **35**, 2355.

Christenson, T.R., Guckel, H., Skrobis, K.J., and Klein, J. (1992). Preliminary results for a integrated planer microdynamometer. In *Solid-State Sensors and Actuators Workshop*, pp. 51–4. IEEE, New York.

Crandall, B.C. and Lewis, J. (ed.) (1992). *Nanotechnology: research and perspective*. MIT Press, Cambridge, MA.

Csepregi, L. (1989). Neue Prozesstechniken in der Mikromechanik. In *Mikromechanik* (ed. A. Heuberger), pp. 216–35. Springer, Berlin.

Dario, P. and Cocco, M. (1993). Technologies and applications of microfabricated implantable neural prosthesis. In *Proceedings of the Workshop on Micromachine Technologies and Systems*, pp. 38–45. Isukuba.

de Rooij, N.F. (1990). Application of micromachined sensors. In *Integrated Micro-Motion Systems—Micro Machining, Control and Application* (eds. F. Harashima), pp. 433–62. Elsevier, Tokyo.

Deimel, P.P. (1989). Mikromechanik in der integrierten Optoelektronik. In *Mikromechanik* (ed. A. Heuberger), pp. 432–61. Springer, Berlin.

Douglas, D.A. and Mann, S. (1993). Template mineralization of self-assembled anisotropic lipid microstructures. *Nature*, **364**, 430–3.

Drexler, K.E. (1986). *Engines of creation: the coming era of nanotechnology.* Anchor Press, New York.

Drexler, K.E. (1987). Nanomachinery: atomically precise gears and bearings. In *IEEE Micro Robots and Teleoperators Workshop (Hyannis, MA)* p. 26. IEEE P, New York.

Drexler, K.E. (1992). *Nanosystems: molecular machinery, manufacturing and computation.* Wiley, New York.

Driscoll, R.J., Youngquest, M.G., and Baldeschwerler, J.D. (1990). Atomic-scale imaging of DNA using scanning tunnelling microscopy. *Nature,* **346,** 294–6.

Dubois, F. (1990). Coelioscopic cholecystectomy—preliminary report of 36 cases. *Annals of Surgery,* **211,** 60–2.

Dunlap, D.D. and Bustamante, C. (1989) Images of single-strand nucleic acids by scanning tunnelling microscopy. *Nature,* **342,** 204–6.

Edell, D.J. (1986). A peripheral nerve information transducer for amputees: long-term multichannel recordings from rabbit peripheral nerves. *IEEE Transactions on Biomedical Engineering,* **33,** 203–14.

Egawa, S. and T. Higuchi (1990). Multi-layered electrostatic film actuator. In *IEEE Microelectromechanical Systems (Napa Valley, CA),* pp. 166–7. IEEE, New York.

Egawa, S., Niino, T., and Higuchi, T. (1991). Film actuators: planer, electrostatic surface-drive actuator. In *IEEE Microelectromechanical Systems (Nara, Japan),* pp. 9–14. IEEE, New York.

Ehrfeld, W., Glashauser, W., Münchmeyer, D., and Schelb, W. (1986). Mask making for synchrotron radiation lithography. *Microelectronic Engineering,* **5,** 463–70.

Ehrfeld, W., Bley, P., Götz, F., Hagemann P., *et al.* (1987). Fabrication of microstructures using the LIGA process. In *IEEE Micro Robots and Teleoperators Workshop (Hyannis, MA),* p. 11. IEEE, New York.

Ehrfeld, W., Götz, F., Münchmeyer, D., Schelb, W., *et al.* (1988). LIGA process: sensor construction techniques via X-ray lithography. In *IEEE Solid State Sensors and Actuators Workshop,* pp. 1–4. IEEE, New York.

Eigler, D. and Schweizer, E. (1990). Positioning single atoms with a scanning tunnelling microscope. *Nature,* **344,** 524–6.

Elwenspoek, M., Blom, F.R., Bouwstra, S., and Lammerink, T.S.J., *et al.* (1989). Transduction mechanisms and their applications in micromechanical devices. In *IEEE Microelectromechanical Systems (Salt Lake City, UT),* pp. 126–32, IEEE, New York.

Emoto, F., Gamo, K., Nambe, K., Samoto, R., *et al.* (1985). Nanometer structure fabrication attained by a nanometer e-beam lithography system. *Microelectronic Engineering,* **3,** 17–24.

Esashi, M. (1994). Sensor for measuring acceleration. In *Mechanical sensors* (ed. N.F. de Rooij), pp. 331–58. VCH, Amsterdam.

Esashi, M., Shoji, S., and Nakano, A. (1989). Normally close microvalve and micropump fabricated on a silicon wafer. In *IEEE Microelectromechanical Systems (Salt Lake City, UT),* pp. 29–34. IEEE, New York.

Fan, L., Tai, Y. and Muller, R.S. (1988). Integrated movable micromechanical structures for sensors and actuators. *IEEE Transactions on Electron Devices,* **35,** 724–30.

Feynman, R. (1961). There's plenty of room at the bottom. In *Miniaturization* (ed. H. D. Gilbert). Reinhold, New York.

Flynn, A.M., Brooks, R.A., Wells III, W.M., and Barrett, D.S. (1989). The world's largest one cubic inch robot. In *IEEE Microelectromechanical Systems (Salt Lake City, UT)* pp. 98–101. IEEE, New York.

Foster, J., Frommer, J., and Arnett, P. (1988). Molecular manipulation using a tunneling microscope. *Nature*, **331**, 324–6.

Fujimasa, I. (1992). Future medical applications of microsystem technologies. In *Microsystem technologies 92 (Berlin)* pp. 43–56. VDE-Verlag, Berlin.

Fujimasa, I. (1994). *Invisible machines*. Ohm-sha, Tokyo.

Fujimasa, I. and Nakjima, N., *et al.* (1989). *Research report on micromachine technology*, p. 152. Tokyo, Technova.

Fujimasa, I., Chinzei, T., Imachi, and K., Matuura, H., *et al.* (1992). Development of integrated actuators using vibrational energy in the environment. *Micromachine*, **5**, 21–6.

Fujita, H. and Omodaka, A. (1987). Electrostatic actuators for micromechatronics. In *IEEE Micro Robots and Teleoperators Workshop (Hyannis, MA)* p. 14. IEEE, New York.

Gabriel, K., Jarvis, J., and Trimmer, W. (ed.) (1988). Small machines, large opportunities: a report on the emerging field of microdynamics. In *The NSF Workshop on Microelectromechanical Systems Research*.

Gabriel, K.J., Behi F., Mahadevan, R., and Mehregany, M. (1990). *In situ* friction and wear measurements in integrated polysilicon mechanisms. *Sensors and Actuators*, **A21-A23**, 180.

Gass, V., van der Schoot, B.H., and de Rooij, N.F. (1993). Nanofluid handlings by micro-flow-sensor based on drag force measurements. In *IEEE Microelectromechanical Systems (Fort Lauderdale, FL)*, pp. 167–72. IEEE, New York.

Gass, V., van der Schoot, B.H., Jeanneret, S., and de Rooij, N.F. (1994). Micro-torque sensor based on differential force measurement. In *IEEE Microelectromechanical Systems (Oiso, Japan)*, pp. 241–4. IEEE, New York.

Gianchandani, Y. and Najafi, K. (1991). In *Digest of the IEEE International Electron Devices Meeting*.

Goldstein, L.S.B. and Vale, R.D. (1991). A brave new world for dynein. *Nature*, **352**, 569–70.

Gruntzig, A. and Hopff, H. (1974). Perkutane Recanalization chronischer arterieller Verschlusse mit einem neuen Dilatation Katheter. *Deutsche Medicinische Wochenschrift*, **99**, 2502–5.

Guckel, H., Strobis, K.J., Christenson, and T.R., Klein, J., *et al.* (1991). Fabrication of assembled micromechanical components via deep x-ray lithography. In *IEEE microelectromechanical systems (Nara, Japan)*, pp. 74–9. IEEE, New York.

Guckel, H., Christenson, T.R., Skrobis, K.J., and Jung, T.S. *et al.* (1993). A first functional current exited planer rotational magnetic micromotor. In *IEEE Microelectromechanical Systems (Fort Lauderdale, FL)*, pp. 7–11. IEEE New York.

Guldberg, J., Nathanson, H.C., Balthis, D.S., and Jensen, A.S. (1975). An aluminium/SiO_2 silicon on sapphire light valve matrix for projection displays. *Applied Physical Letters*, **26**, 391.

Harada, Y., Sakurada, K., Aoki, T., and Thomas, D.D., *et al.* (1990). Mechanochemical coupling in actomyosin energy transduction studied by *in vitro* movement assay. *Molecular Biology*, **216**, 49–68.

Harashima, F. (ed.) (1990). *Intergrated micro-motion systems: micromachining, control and applications*. Elsevier, Amsterdam

Hashimoto, D. (1992). Stereo-laparoscope developed by Shinko-koki Co. In *13th Annual Meeting of Japan Society for Laser Medicine (Tokyo)*. Personal communication.

Hatamura, Y. and H. Morishita (1990). Direct coupling system between nanometer world and human world. In *IEEE Microelectromechanical Systems (Napa Valley, CA)* pp. 203–8. IEEE, New York.

Hatamura, Y., Nakao, M., Sato, T., and Koyano, K., *et al.* (1994). Construction of 3-D micro structure by multi-face FAB, co-focus rotational robot and various mechanical tools. In *IEEE Microelectromechanical Systems (Oiso, Japan)*, pp. 297–302. IEEE New York.

Hayashi, I. (1990). Reports of 'Hill Climb Micromaze Contest'. *Seimitsu Kougakukai-shi (J. of Precision Engineering)*, **56**, 2189.

Hayashi, T. (1993). Feasible trends of new actuators. *Proceedings of the New Actuator and Sensor-fusion Symposium 93*. Nihon-nouritsu Kyokai, N1: 1–9, Tokyo.

Hayashi, T. (1985). Characteristics of an actuator used for precise feeding mechanisms. *Proceedings of the Annual Spring Meeting of the Japanese Society for Precision Engineering*, pp. 185–9.

Heller, M.J. (1996). An active microelectronics device for multiplex DNA analysis. *IEEE Engineering in Medicine and Biology*, **15**(2), 100–4.

Herzenberg, L.A., Sweet, R.G., and Herzenberg, L.A. (1976). Fluorescence-activated cell sorting. *Scientific American*, **234**, (3), 108–15

Hetke, J.F., Najafi, K., and Wise, K.D. (1991). Flexible silicon interconnects for microelectromechanical systems. In *Proceedings of the 6th International Conference on Solid-State Sensors and Actuators (Transducers '91) (San Francisco, CA)*, pp. 1764–767

Hetke, J.F., Lund, J.L., Najafi, K., and Wise, K.D., *et al.* (1994). Silicon ribbon cables for chronically implantable microelectrode arrays. *IEEE Transactions on Biomedical Engineering*, **41**, 314–21.

Heuberger, A. (1986). X-ray lithography. *Microelectronic Engineering*, **5**, 3–38.

Heuberger, A. (ed.) (1989). *Mikromechanik*. Springer, Berlin.

Heuberger, A. and Betz H. (1983). X-ray-lithography with synchrotron radiation. In *Solid state devices* (ed. A. Goetzberger and M. Zerbst). Verlag Chemie, Weinheim.

Higuchi, T., Hojjat, Y., and Watanabe, M. (1987). Micro actuators using recoil of an ejected mass. In *IEEE Micro Robots and Teleoperators Workshop (Hyannis, MA)* p. 21. IEEE, New York.

Higuchi, T., Yamagata, Y., Furutani, K., and Kudoh, K. (1990). Precise positioning mechanism utilizing rapid deformations of piezoelectric elements. In *IEEE Microelectromechanical Systems (Napa Valley, CA)* pp. 222–6. IEEE New York.

Hirokawa, N. (1991). Molecular architecture and dynamics of the neuronal cyto skeleton. In *The neuronal cyto steleton*, pp. 5–74. Wiley-Liss, New York.

Hirokawa, N., Pfister, K.K., Yorifuji, H., and Wagner, M.C., *et al.* (1989). *Cell*, **56**, 867–78.

Hirose, S. (1987). *Biological mechanical engineering: principle of soft robots* Kogyocyosakai, Tokyo. (In Japanese.)

Hoogerwerf, A.C. and Wise, K.D. (1991). A three dimensional neural recording array. In *Digest of the IEEE International Conference on Solid-State Sensors and Actuators (San Francisco, CA)*, pp. 120–3.

Howe, R.T., Muller, R.S., Gabriel, K.J., and Trimmer, W.S.N. (1990). Silicon micromechanics: sensor and actuators on a chip. *IEEE Spectrum* July 1999, 29–35.

Huber, H.-L. and Betz, H. (1989). Tiefenlithographie und Abformtechnik. In *Mikromechanik* (ed. A. Heuberger), pp. 236–64. Springer, Berlin.

Huxley, A.F. (1957). Muscle structure and theories of contraction. *Progress in Biophysics and Biophysical Chemistry* **7**, 255–318.

Igarashi, I. (1990). Microsensors and its applications. In *Integrated micro-motion systems—micromachining, control and applications* (ed. F. Harashima), pp. 201–20. Elsevier, Tokyo.

Iijima, S. and Ichihashi, T. (1993). Single-shell carbon nanotubes of 1 nm diameter. *Nature*, **363**, 603–6.

Ikuta, K. (1988*a*). Shape memory alloy servo actuator system with electric resistance feedback and application for active endoscope. In *Proceedings of the IEEE International Conference on Robotics and Automation*, pp. 427–30.

Ikuta, K. (1988*b*). Applications of the shape memory alloy actuators to micromechanisms. *Journal of the Japan Society of Precision Engineering*, **54**, 1656–61.

Ikuta, K. and Hirowatari, K. (1993). Real three-dimensional micro fabrication using stereo lithography and metal molding. In *IEEE Microelectromechanical Systems (Fort Lauderdale, FL)*, pp. 42–7. IEEE, New York.

Ikuta, K., Fujita, H., Ikeda, M., and Yamashita, S. (1990). Crystallographic analysis of Ti–Ni shape memory alloy thin film for micro actuator. In *IEEE Microelectromechanical Systems (Napa Valley, CA)*, pp. 38–9. IEEE, New York.

Imachi, K., Mabuchi, K., Chinzei, T., and Abe, Y., *et al.* (1993). The jellyfish valve: a polymer membrane valve for the artificial heart. In *Heart replacement: artificial heart 4* (ed. T. Akutsu and H. Koyanagi), pp. 41–4. Springer, Tokyo.

Ishihara, K., Tanouchi, J., and Kitabatake, A., *et al.* (1991). Noninvasive and precise motion detection for micromachines using high-speed digital subtraction echography (high-speed DSE). In *IEEE Microelectromechanical Systems (Nara, Japan)*, pp. 176–81. IEEE, New York.

Ishijima, A., Doi, T., Sakurada, K., and Yanagida, T. (1991). Sub-piconewton force fluctuations of actomyosin *in vitro*. *Nature*, **352**, 301–6.

Israelachvili, J.N., *et al.* (1990). Liquid dynamics in molecularly thin films. *Journal of Physics: Condensed Matter*, **2**.

Jacobsen, S. (1987). Micro electro-mechanical systems (MEMS). In *IEEE Micro Robots and Teleoperators Workshop (Hyannis, MA)*, p. 12. IEEE, New York.

Jacobsen, S.C., Price, R.H., Wood, J.E., Rytting, T.H., *et al.* (1989). The wobble motor: an electrostatic, planetary-armature microactuation. In *IEEE Microelectromechanical Systems (Salt Lake City, UT)*, pp. 17–24. IEEE, New York.

Ji, J. (1990). A scaled electrically-configurable CMOS multichannel intracortical recording array. Thesis, University of Michigan.

Ji, J., Najafi, K., and Wise, K.D. (1991). A low-noise demultiplexing system for active multichannel microelectrode arrays. *IEEE Transactions on Biomedical Engineering*, **38**, 75–81.

Jimbo, Y. and Kawana, A. (1992). Electrical stimulation and recording from cultured neurons using a planar electrode array. *Bioelectrochemistry and Bioenergetics*, **29**, 193–204.

Jono, K., Hashimoto, M., and Esashi, M. (1994). Electrostatic servo system for multi-axis accelerometer. In *IEEE Microelectromechanical Systems (Oiso, Japan)*, pp. 254–6. IEEE, New York.

Kabata, H., Kurosawa, O., Arai, I., and Washizu, M., *et al.* (1993). Visualization of single molecules of RNA polymerase sliding along DNA. *Science*, **262**, 1561–3.

Kami-ike, N., Kudo, S., Magariyama, Y., and Aizawa, S., *et al.* (1991*a*). Characteristics of an ultra-small biomotor. In *IEEE Microelectromechanical Systems (Nara, Japan)* pp. 245–6. IEEE, New York.

Kami-ike, N., Kudo, S., and Hotani, H. (1991*b*). Rapid charges in flagellar rotation induced by external electric pulses. *Biophysical Journal*, **60**, 1350–5.

Kaneko, R. (1991). Microtribology related MEMS. In *IEEE Microelectromechanical Systems (Nara, Japan)*, pp. 1–8. IEEE, New York.

Kato, A., Yoshimine, T., Hayakawa, T., and Tomita, Y., *et al.* (1991). A frameless navigational system for computer-assisted neurosurgery. *Journal of Neurosurgery*, **74**, 845–9.

Kim, Y., Katurai, M., and Fujita, H. (1989). A proposal for a superconducting actuator using the Meissner effect. In *IEEE Microelectromechanical Systems (Salt Lake City, UT)*, pp. 107–22. IEEE, New York.

Kirshner, J.M. (ed.) (1966). *Fluid amplifiers*. McGraw-Hill, New York.

Kishino, A. and Yanagida, T. (1988). *Nature*, **334**, 74–6.

Kobayashi, K. (1988). Laser assist etching. *Seimitsu-kougakukai Zasshi* (Journal of the Japanese Society of Precision Machining), **54**, 1673–7.

Komine, H., Takahashi, O., Saito, H., Togawa, T., and Tsuchiya, K. (1994). Automatic blood sampler aimed at blood sucking mechanism of mosquito. *Japanese Journal of Medical Electronics and Biological Engineering*, **32** (suppl), 283.

Kong, L.C., Orr, B.G., and Wise, K.D. (1990). In *Digest of the 1990 IEEE Solid-State Sensor and Actuator Workshop*, p. 28. IEEE, New York.

Kovacs, G.T.A. (1991). Regeneration microelectrode arrays for direct interface to nerves. In *Technical Digest of the International Conference on Solid-State Sensors and Actuators (Transducers '91) (San Francisco, CA)*, pp. 116–19.

Kovacs, G.T.A., Stormet, C.W., and Rosen, J.M. (1992). Regeneration microelectrode array for peripheral nerve recording and stimulation. *IEEE Transactions on Biomedical Engineering*, **39**, 893–902.

Kubo, Y., Shimoyama, I., and Miura, H. (1993). In *Proceedings of the IEEE International Conference on Robotics and Automation.*

Lang, J.H., Schlecht, M.F., and Howe, R.T. (1987). Electric micromotors: electro-mechanical characteristics. In *IEEE Micro Robots and Teleoperators Workshop (Hyannis, MA)* p. 13. IEEE, New York.

Langer, R. and Vacanti, J.P. (1993). Tissue engineering. *Science*, **260**, 920–6.

Leob, G.E., Marks, W.B., and Beatty, P.G. (1977). Analysis and microelectronic design of tublar electrode arrays intended for chronic, multiple single-unit recording from captured nerve fibers. *Medical and Biological Engineering and Computation*, **15**, 195–201.

Mann, S. (1993). Molecular tectonics in biomineralization and biomimetic materials chemistry. *Nature*, **365**, 499–505.

Mannerd, A., Stein, R.B., and Charles, D. (1974). Regeneration electrode units: implant for recording from peripheral nerve fibers in freely moving animals. *Science*, **183**, 547–9.

Marks, A.F. (1969). Bullfrog nerve regeneration into porous implants. *Anatomical Record*, **163**, 266.

Masaki, T., Kawata, K., and Masuzawa, T. (1990). Micro electrodischarge machining and its applications. In *IEEE Microelectromechanical Systems (Napa Valley, CA)*, pp. 21–6. IEEE, New York.

Matsuda, T., Inoue, K., and Sugawara, T. (1989). Micro-patterning of cultured cells. *Transactions of the American Society for Artificial Internal Organs.*

Matsuo, T. (ed.) (1988). *Technical Digest of the 7th Sensor Symposium*. The Institute of Electrical Engineers of Japan, Tokyo.

Matsuo, T. (ed.) (1989). *Technical Digest of the 8th Sensor Symposium*. The Institute of Electrical Engineers of Japan, Tokyo.

Matsuura, H. (1993). Research of artificial muscles. Thesis, Graduate School of Medicine, University of Tokyo, pp. 125.

May, G., Shamma, S.A., and White, R.L. (1979). A tantalum-on-sapphire microelectrode array. *IEEE Transactions on Electron Devices*, **29**, 1932–9.

Meheregany, M., Gabriel, K.J., and Trimmer, W.S.N. (1988). Integrated fabrication of polysilicon mechanisms. *IEEE Transactions on Electron Devices*, **35**, 719–23.

Meheregany, M., Nagarkar, P., Senturia, S.D., and Lang, J.H. (1990). Operation of microfabricated harmonic and ordinary side-drive motors. In *IEEE Microelectromechanical Systems (Napa Valley, CA)*, pp. 1–8. IEEE, New York.

Menz, W., Bacher, W., Harmening, M., and Michel, A. (1991). The LIGA technique: a novel concept for microstructures and the combination with Si technologies by injection molding. In *IEEE Microelectromechanical Systems (Nara, Japan)*, pp. 69–73. New York.

Middelhoek, S. (1989). Innovative silicon sensors for radiant, mechanical, thermal, magnetic and chemical signals. In *Integrated Micro Motion Systems—Micromachining, Control and Applications* (ed. F. Harashima), pp. 21–50. Elsevier, Tokyo.

Mikuriya, Y., Matsuzaki, K., and Matsuo, T. (1993). Fabrication and evaluation of micromotors. In *IARP Workshop on Microsystems and Robotics (Karlsruhe, Germany)*.

Miyake, S. (1992). Technological trend and its break through for anti-wearness coating.) *Tribologist*, **37**, 715.

Miyake, S. and Kaneko, R. (1992). Micotribological properties and potential applications of hard, lubricating coatings. *Thin Solid Films*, **212**, 256.

Miyake, S., Watanabe, S., Murakawa, M., Kaneko, R., *et al.* (1992). Tribological study of cubic boron nitride film. *Thin Solid Films*, **212**, 262.

Miyamoto, T., Miyake, S., and Laneko, R. (1993). Wear resistance of C+-implanted, silicon investigated by scanning probe microscope. *Wear*, **162–4**, 733.

Mizoguchi, H., Ando, M., Mizuno, T., and Takagi, T., *et al.* (1992). Design and fabrication of light driven micropump. In *IEEE Microelectromechanical Systems (Travemünde, Germany)*, pp. 31–6. IEEE, New York.

Motokawa, T. (1993). *Time of an elephant and a mouse*. Cynonkouron-sya, Tokyo.

Muller, R.S. (1987). From ICs to microstructures: materials and technologies. In *IEEE Micro Robots and Teleoperators Workshop, (Hyannis, MA)*, p. 2. IEEE, New York.

Muller, R.S., Howe, R.T. Senturia, S.D., and Smith, R.L., *et al.* (ed.) (1991). *Microsensors*. IEEE, New York.

Myler, R.K., Cumberland, P.A., and Clark, D.A. (1987). High and low power thermal laser angioplasty for total occulusion and restenoses in man. *Circulation*, **76**, Suppl. IV, 238.

Nagamori, S. (1990). Design and machining of micromachines with UV-hardening-polymer. *Kikai Sekkei* (Machine Designing), **34**, (15), 50–5.

Najafi, K. (1994). Solid-state microsensors for cortical nerve recordings. *IEEE Engineering in Medicine and Biology*, **13**, 375–87.

Najafi, K. and Hetke, J.F. (1990). Strength characterization of silicon microprobes in neurophysiological tissues. *IEEE Transactions on Biomedical Engineering*, **37**, 474–81.

Najafi, K. and Wise, K.D. (1986). An implantable multielectrode recording array with on-chip signal processing. *IEEE Journal of Solid-State Circuits*, **21**, 1035–44.

Najafi, K., Wise, K.D., and Mochizuki, T. (1985). A high-yield IC-compatible multichannel recording array. *IEEE Transactions on Electron Devices*, **32**, 1206–11.

Nakajima, N., Ogawa, K., and Fujimasa, I. (1989). Study on micro engines: miniaturizing Stirling engines for actuators. *Sensors and Actuators*, **20**, 75–82.

Neher, E. and Sakmann, B. (1976). *Journal of Physiology*, **253**, 705.

Overney, R.M., Meyer, E., Frommer, J., Brodbeck, D., *et al.* (1992). Friction measurements on phase-separated thin films with a modified atomic force microscope. *Nature*, **359**, 133–5.

Orgel, L.E. (1992). Molecular replication. *Nature*, **358**, 203–9.

Ouano, A.C. (1978). A study on the dissolution rate of irradiated polymethylmethacrylate. *Polymer Engineering and Science*, **18**, 306–13.

Pelka, J. and Weigmann, W. (1989). Einsatz von Ionentechniken. In *Mikromechanik* (ed. A. Heuberger), pp. 171–212. Springer, Berlin.

Perissat, J. (1990). Gallstones: laparoscopic treatment—cholecystectomy cholecystostomy and lithotripsy. *Surgical Endoscopy*, **4**, 1–5.

Petersen, K.E. (1982). Silicon as a mechanical material. *Proceedings of the IEEE*, **70**, 420–57.

Petersen, K. (1987). The silicon micromechanics foundry. In *IEEE Micro Robots and Teleoperators Workshop* (*Hyannis, MA*), p. 1. IEEE, New York.

Pethica, J.B. (1988). Scanning tunnelling microscopes: atomic-scale engineering. *Nature*, **331**, 301.

Petzold, H.-C. (1989). Laserinduzierte Prozesse. In *Mikromechanik* (ed. A. Heuberger), pp. 265–328. Springer, Berlin.

Quate, C.F. (1990). Imaging with the tunneling and force microscopes. In *IEEE Microelectromechanical Systems* (*Napa Valley, California*), pp. 188–9. IEEE, New York.

Rachine, G., Luthier, R., and de Rooij, N.F. (1993). Hybrid ultrasonic micromachined motors. In *IEEE Microelectromechanical Systems* (*Fort Lauderdale, FL*), pp. 128–38. IEEE, New York.

Reddick, E.J. and Olsen, D.O. (1989). Laparoscopic laser cholecystectomy. *Surgical Endoscopy*, **3**, 131–3.

Reichl, H. (1989). Mikromechanik und Chipverbindungstechnik. In *Mikromechanik* (ed. A. Heuberger), pp. 462–86. Springer, Berlin.

Reichl, H. (ed.) (1992). *Micro system technologies 92.* VDE, Berlin.

Reithmuller, W. and Benecke, W. (1988). Thermally excited silicon microactuators. *IEEE Transactions on Electron Devices*, **35**, 758–63.

Robbins, H. and Schwartz, B. (1958). Chemical etching of silicon, II. the system HF, HNO_3, $HC_2H_3O_2$. *Journal of the Electrochemical Society*, **106**, 505.

Robertson, J.K. and Wise, K.D. (1994). A nested electrostatically actuated microvalve for an integrated microflow controller. In *IEEE Microelectromechanical Systems* (*Oiso, Japan*), pp. 11–12. IEEE, New York.

Rosen, J.M., Kovacs, G.T.A., Stephanides, M., Marshall, D., *et al.* (1987). The development of a microelectronic axon processor silicon ship neuroprosthesis. In *Proc. Int. Conf. Ass. Adv. Rehab. Tech.* (*San Jose, CA*), pp. 675–7.

Roylance, L.M. and Angell, J.B. (1979). A batch-fabricated silicon accelerometer. *IEEE Transactions on Electron Devices*, **26**, 1911–17.

Runyan, W.R., Alexander, E.G., and Craig, S.E. Jr (1967). Behavior of large-scale surface pertubations during silicon epitaxial growth. *Journal of the Electrochemical Society*, **114**, 1154.

Sakmann, B. (1992). Elementary steps in synaptic transmission revealed by currents through single ion channels. *Science*, **256**, 503–12.

Sanborn, T.A., Curberland, D.C., and Greenfield, A.J. (1988). Percutaneous laser thermal angioplasty initial results and 1 year follow-up in 129 femoropopliteal lesions. *Radiology*, **168**, 121–5.

Schultz, J.S. (1991). Biosensors. *Scientific American*, **265**, (2), 48–55.

Seidel, H. (1989). Nasschemisch Tiefenätztechnik. In *Mikromechanik* (ed. A. Heuberger), pp. 125–70. Springer, Berlin.

Senturia, S.D. (1987). Can we design microbotic devices without knowing the mechanical properties of materials? In *IEEE Micro-Robots and Teleoperators Workshop* (*Hyannis, MA.*), p. 3. IEEE, New York.

Skrzypek, J. and Keller, E. (1975). Manufacture of metal microelectrode with the scanning electron microscope. *IEEE Transactions on Biomedical Engineering*, **22**, 435–8.

Sotobayashi, H., Asmussen, F., Thimm, K., and Schnabel, W., *et al.* (1982). Degradation of polymethylmethacrylate by synchrotron radiation. *Polymer Bulletin*, **7**, 95–101.

Sugiyama, S., Suzuki, T., Kawahata, K., and Simaoka, K., *et al.* (1986). Micro-diaphragm pressure sensor. In *IEEE International Electron Devices Meeting*, pp. 184–7. IEEE, New York.

Suzuki, S. *et al.* (1990). Semiconductor capacitance-type accelerometer with PWM electrostatic servo technique. *Sensors and Actuators*, **A21–A23**, 316–19.

Tai, Y.C. and Muller, R.S. (1990). Friction study of IC-processed micromotors. *Sensor and Actuators*, **A21-A23**, 180.

Tai, Y.C., Fan, L.S., and Muller, R.S. (1989). IC-processed micro-motors: design, technology, and testing. In *IEEE Microelectromechanical Systems* (*Salt Lake City, UT*) pp. 1–6. IEEE, New York.

Takagi, T. and Nakajima, N. (1993). Photoforming applied to fine machining. In *IEEE Microelectromechanical Systems* (*Fort Lauderdale, Florida*), pp. 173–8. IEEE, New York.

Takahashi, K. and Matsuo, T. (1984). Integration of multi-micro electrode and interface circuits by silicon planer and three-dimensional fabrication technology. *Sensors and Actuators*, **5**, 89–99.

Tan, W., Shi, Z.Y., Smith, S., Birnbaum, D., *et al.* (1992). Submicrometer intracellular chemical optical fiber sensors. *Science*, **258**, 778–81.

Tang, W.C., Nguyen, T.H., and Howe, R.T. (1989). Laterally driven polysilicon resonant microstructures. *Proc. IEEEMEMS*, 53–9.

Taylor, R.H., Lavallée, S., Burdea, G.C., and Mösges, R. (ed.) (1986). *Computer-integrated surgery*. MIT Press, Massachusetts.

Terry, S.C., Jerman, J.H., and Angell, J.B. (1979). A gas chromatograph air analyzer fabricated on a silicon wafer. *IEEE Transactions on Electron Devices*, **26**, 1880–6.

Travis, J. (1992). Making materals that are good to the last drop. *Science*, **258**, 1307.

Trimmer, W. (1990). Micromechanical systems. In *Integrated micro-motion systems— micromachining, control and applications* (ed. F. Harashima), pp. 1–19. Elsevier, Amsterdam.

Turner, D.R. (1958). Electropolishing silicon in hydrofluoric acid solutions. *Journal of the Electrochemical Society*, **126**, 1406.

Waggener, H.A., Kragness, R.C., and Taylor, A.L. (1967). *Electronics*, **40**, 274.

Vale, R.D. (1993). Measuring single protein motors at work. *Science*, **260**, 169–70.

Wagner, B., Kreutzer, M., and Benecke, W. (1992). Linear and rotational magnetic micromotors fabricated using silicon technology. In *IEEE Microelectromechanical Systems* (*Travemünde, Germany*), pp. 183–9. IEEE, New York.

Washizu, M. (1992). Manipulation of biological objects in micromachine structures. In *IEEE Microelectromechanical systems* (*Travemunde, Germany*) pp. 196–20. IEEE, New York.

Wickramasinghe, K. (1991). Scanned probe microscopy and manipulation. In *Second Conference on Molecular Nanotechnology (Palo Alto, CA)*.

Wilson, A.D. (1985). X-ray lithography: can it be justified? *Proceedings of the SPIE*, **537**, 85–101.

Wise, K.D. and Angell, J.B. (1975). A low-capacitance multielectrode probe for use in extracellular neurophysiology. *IEEE Transactions on Biomedical Engineering*, **22**, 212–19.

Wise, K.D. and Najafi, K. (1991). Microfabrication techniques for integrated sensors and microsystems. *Science*, **254**, 1335–42.

Yamaguchi, M., Kawamura, S., Minami, K., and Esashi, M. (1993). Distributed electrostatic micro actuator. In *IEEE Microelectromechanical Systems (Fort Lauderdale, FL)*, pp. 18–23. IEEE, New York.

Yanagida, T. (1989). Observation of molecular motion for gliding muscle. *Frontiers in Biophysics: Protein, Muscle Contraction, and Brain and Neurons* (ed. Japanese Society of Biophysics, Baifukann, Tokyo, Japan), pp. 121–33.

Yauo, M., Yamanuoto, Y., and Shimizu, H. (1982). *Nature*, **229**, 557–9.

Ziaie, B., VonArx, J., Nardin, M., and Najafi, K. (1993). A hermetic packaging technology with multiple feedthroughs for integrated sensors and actuators. In *International Conference on Solid-State Sensors and Actuators (Transducers '93) (Yokohama, Japan)*, pp. 266–9.

Zmood, R.B., Bates, I., and Sood, D. (1993). *Micro engineering and micro machines: a new threshold for Australian industry*. Office of the Chief Scientist, Department of the Prime Minister and Cabinet, Australia.

Index